The
Reference Shelf ®

Gene Editing & Genetic Engineering

The Reference Shelf
Volume 95 • Number 5
H.W. Wilson
a Division of EBSCO Information Services, Inc.

Published by
GREY HOUSE PUBLISHING
Amenia, New York
2023

The Reference Shelf

Cover photo: ipopba/iStock

The books in this series contain reprints of articles, excerpts from books, addresses on current issues, and studies of social trends in the United States and other countries. There are six separately bound numbers in each volume, all of which are usually published in the same calendar year. Numbers one through five are each devoted to a single subject, providing background information and discussion from various points of view and concluding with an index and comprehensive bibliography that lists books, pamphlets, and articles on the subject. The final number of each volume is a collection of recent speeches. Books in the series may be purchased individually or on subscription.

∞ The paper used in these volumes conforms to the American National Standard for Permanence of Paper for Printed Library Materials, Z39.48 1992 (R2009).

Publisher's Cataloging-in-Publication Data
(Prepared by Parlew Associates, LLC)

Names: Grey House Publishing, Inc., compiler.
Title: Gene editing & genetic engineering / [compiled by Grey House Publishing].
Other Titles: Reference shelf ; v. 95, no. 5.
Description: Amenia, NY : Grey House Publishing, 2023. | Includes bibliographic references and index. | Includes b&w photos and illustrations.
Identifiers: ISBN 9781637004982 (v.95, no. 5) | ISBN 9781637004937 (volume set)
Subjects: LCSH: Genetic engineering. | Genetic engineering—Moral and ethical aspects. | Gene editing. | Gene therapy. | BISAC: SCIENCE / Life Sciences / Genetics & Genomics. | TECHNOLOGY & ENGINEERING / Biomedical. | MEDICAL / Genetics.
Classification: LCC QH442.G46 2023 | DDC 660.65--dc23

Printed in Canada

Contents

3

Genetic Medicine

4

Controversial Technology

5

Popular Genetics

Preface

Mastering the Gene

Gene Editing and Genetic Engineering

Genetics is the study of "genes," which are molecules that essentially contain instructions for how to make other molecules. During reproduction, these instructional molecules are utilized by the machinery within an oocyte, or "egg cell" to create the various proteins, lipids, and carbohydrates that make up the body of an organism. Because the genes contain the instructions, genes determine what an organism will look like and other characteristics that the organism might have. While genes play a major role, however, they are only part of the story. Every animal and plant has traits that are shaped by both internal (genetic) and external (environmental) factors and so genes are not the only factor determining what an organism looks like or how its body works (or fails to work). Further, multiple genes are typically involved in any trait. Baldness, for instance, is not related to a single gene that "causes" baldness but is the result of interacting genetic factors inherited through both parents and influenced further by environmental factors.[1]

Genetic engineering or "genetic modification" is the process of using technology to alter the genetic makeup of an organism. This can be done before the organism is born, during the cellular division stage, or it can be done when an organism has already been born and is at pretty much any stage in its growth. Genetic engineering can be used to alter a gene, to remove a gene, or even to add a gene to an organism's genome. In some cases, this can be used to address a genetic disorder or disease, by eliminating a malfunctioning gene or genetic sequence or changing a gene to change the way that the gene is expressed. In other cases, genetic engineering can be used to enhance or to alter an organism's form or function, and this is where genetic modification tends to become more controversial. Many Americans feel that genetic modification might sometimes be acceptable in situations where modification is used to help someone struggling with a disease but are skeptical about altering genomes to enhance disease or to change an individual's genetic structure in ways not related to treating disease.[2]

The terminology and techniques surrounding genetic modification can be difficult to understand for the lay public and the rapidly evolving field is also changing so fast that it can be difficult for people to keep up with the field. The terms "genetic engineering" and "genetic modification" can be used to refer to any kind of genetic alteration, but there are a number of subfields and more specific techniques as well. Terms like "negative genetic engineering," or "therapeutic genetic engineering" refer to a field generally known as "gene therapy," where the goal is to use genetic modification to address a problem that a person is having, such as a disease or physical conditions that is related to genetics. The terms "positive genetic engineering" or "genetic enhancement" refers to the use of genetic engineering to alter a genome in order to enhance an organism's characteristics. It has been suggested, for instance, that genetic engineering could be used to provide humans with professional-athlete level physical capabilities, and such a project would be an example of genetic enhancement. In addition, scientists have created pigs that produce less hazardous waste, and this would be considered "genetic enhancement" rather than "genetic therapy."[3]

Genetic modification can use two different kinds of cells: reproductive cells and somatic cells. The term "somatic cells" refers to pretty much all the cells that exist in the body, making up the muscles, organs, and other tissues. Genes in these cells can be changed in order to perform gene therapy or genetic enhancement and genetic modification of somatic cells changes a person's body and gene expression in the immediate sense. Reproductive cells, like sperm cells, eggs cells, and embryotic cells, can be treated by genetic modification in order to pass on genetic variation to an organism's offspring. For instance, if physicians know that a parent will pass on a genetic disease, it might be possible to modify the individual's reproductive cells so that the individual does not pass on the gene or genes connected to the genetic disorder in question. This is a kind of genetic modification, attempting to prevent the reproductive transmission of genetic conditions or characteristics.

There are several different ways to change the genes within a person's body. First, deoxyribonucleic acid (DNA) within a cell can be replaced by a different or altered DNA sequence, a technique known as "homologous replacement," or the genes can be altered within a person's body.[4] One of the most famous techniques of altering genes is through the use of Clustered Regularly Interspaced Short Palindromic Repeats, but more often called "CRISPR," or "crisper." This tool can be "programmed" to selectively change or eliminate segments of a genetic code and so CRISPR is essentially a gene-editing tool. Another CRISPR system has been developed to alter and edit the genetic information in ribonucleic acid (RNA) molecules.[5]

Once DNA is altered, or new genetic information has been introduced, there are several additional options for how to spread this genetic information through a body or to change additional cells. If gene editing or gene replacement is used on reproductive cells, the introduced changes will be spread to all resulting cells, and this is how genetic engineers can use altered reproductive cells to create new versions of a species. When altering somatic cells, however, it is impractical to manually alter every cell in the body or even within a single organ. To get around this difficulty, it is possible to use a virus as a vector to deliver altered DNA to different cells by essentially spreading genetic alteration as an "infection."

Genetic engineering offers the possibility of breakthrough medical treatments capable of addressing many long-term and largely incurable health issues. Gene therapy and gene editing can aid physicians in addressing genetic malfunctions, or even to replace or fix genes linked to genetic conditions. Genetic medicine empowers scientists and physicians to get at the root of genetic abnormalities and dysfunction and recent breakthroughs also open up the possibility of providing genetic treatments in vitro, eliminating life-threatening or deleterious genetic conditions before they are expressed.

In addition to the potential medical advances, scientists are looking into using genetic techniques for a wide variety of other applications. In the 2010s and 2020s, for instance, scientists have been working on engineering microorganisms that can help humanity address climate change by reducing carbon emissions and helping to address pollution and waste. Scientists have even experimented with making plants that are more efficient at eliminating pollution and pigs that produce less environmentally hazardous waste during digestion. These are only some of the many ways that scientists are attempting to use emerging genetic science to address pressing health and welfare issues arising from climate change.

While many welcome the potential advancements of genetic engineering, the development of this technology has also been controversial, drawing on fears of scientific misconduct and fears that greed will lead to the use of genetic engineering in ways that pose a threat to human welfare.

The Genetics Controversies

In the ancient world, humanity began trying to shape the future of the human species by selectively reproducing different individuals who had what were perceived as beneficial or advanced characteristics. In many monarchic societies, for instance, members of royal families were expected to reproduce only with members of other royal families, both to create political alliances, and to preserve the imaginary God-given leadership ability that was used by royalty and their supporters to justify the authoritarian rule of a lineage or family. Of course, because human in the ancient world had little understanding of genetics, royal marriages led to royal lineages becoming increasingly inbred.

The process of controlled breeding is a simple, ancient form of genetic engineering, selecting "traits" that are connected to genes and then breeding animals (or humans) to enhance these trats. The monarchic breeding practices and efforts to control the supposedly divine or "superior" royal families were promoted to the public as a way to maintain some divine vision for humanity's future but hid a malignant purpose. The restricted inbreeding of royal lineages was used by aristocratic families to hoard wealth and to further dominate the masses. Over the centuries that followed, efforts to "shape" the characteristics human species were rarely any more beneficial or benign. The pseudoscience of "eugenics" a racist, white supremacist ideology disguised as a scientific movement, sought to limit the reproduction of undesirable traits, like dark skin color and mental and physical differences seen as "disabilities," in an effort to construct an advanced version of the white race. Eugenics was, for many years in America, used to justify prejudicial and discriminatory laws and to manipulate the education of children; while eugenics was not actually a science, the propagandistic use of pseudoscientific principles is an example of the kind of activity that has generated a fear and mistrust of mainstream science.[6]

As scientists developed the ability to alter genes, humanity came one step closer to potentially being able to eliminate certain kinds of genetic diseases, but is this technology too dangerous and will some authoritarian states use the power to alter human characteristics as an opportunity to elevate their own power and those "like them?" Will this technology essentially homogenize the human species and become a tool for racists and xenophobes to use against their perceived enemies?

While the promise of genetic engineering is undeniable, many Americans and people around the world are skeptical that scientists can achieve the ultimate goals of genetic engineering without humanity suffering from as-yet-unknown consequences. As with the selective breeding of humans discussed above, many fear that malevolent intentions, greed, or simply carelessness could transform the promise of genetic manipulation into another technological peril with which humanity must contend. Further, many feel a sense of personal, emotional revulsion at the idea of genetically engineered organisms, which are seen by some as "unnatural." For those of the religious persuasion, genetic engineering has been compared to "playing God," by interfering in arenas of existence that seen as reserved for divine force. Whether one feels that biotechnology is antireligious, antinature, or simply a potentially dangerous technology that could be misused by those will ill intentions, few arenas of scientific development have been as scrutinized and have inspired as much fear and concern.

It can be argued that humanity does not have a good track record when it comes to manipulating genetics and this is part of the reason that more modern developments in genetic sciences have often been met with skepticism and suspicion, even inspiring protests and legislative efforts to limit or even prohibit certain kinds of genetic science. This has been one of the major themes in the public discussion of genetic editing and engineering, but the public debate, like the science of genetic engineering itself, is rapidly evolving.

Works Used

"Eugenics and Scientific Racism." *National Human Genome Research Institute/ NIH.* www.genome.gov/about-genomics/fact-sheets/Eugenics-and-Scientific-Racism. Accessed Aug. 2023.

"Gene Therapy and Genetic Engineering." *MU School of Medicine.* University of Missouri. medicine.missouri.edu/centers-institutes-labs/health-ethics/faq/gene-therapy#:~:text=The%20distinction%20between%20the%20two,organism%20 beyond%20what%20is%20normal. Accessed Aug. 2023.

"Genetic Engineering." *National Human Genome Research Institute/ NIH.* Aug. 31, 2023. www.genome.gov/genetics-glossary/Genetic-Engineering#:~:text=Genetic%20engineering%20(also%20called%20 genetic,a%20new%20segment%20of%20DNA. Accessed Aug. 2023.

Hallmann, Armin, Annette Rappel, and Manfred Sumper. "Gene Replacement by Homologous Recombination in the Multicellular Green Alga *Volvox carteri*." *PNAS* 94, no. 14, July 8, 1997.

"Questions and Answers about CRISPR." *Broad Institute*. www.broadinstitute.org/what-broad/areas-focus/project-spotlight/questions-and-answers-about-crispr. Accessed Aug. 2023.

"What Is a Gene?" *Medline Plus*. National Library of Medicine. medlineplus.gov/genetics/understanding/basics/gene/. Accessed Aug. 2023.

Notes

1. "What Is a Gene?" *Medline Plus*.
2. "Genetic Engineering," *National Human Genome Research Institute/NIH*.
3. "Gene Therapy and Genetic Engineering," *MU School of Medicine*.
4. Hallmann, Rappel, and Sumper, "Gene Replacement by Homologous Recombination in the Multicellular Green Alga *Volvox carteri*."
5. "Questions and Answers about CRISPR," *Broad Institute*.
6. "Eugenics and Scientific Racism," *National Human Genome Research Institute/NIH*.

1
Genetic Manipulation

Photo by Herb Pilcher, USDA Agricultural Research Service, via Wikimedia.

Genetically manipulated peanut leaves (bottom image) that are protected from damage caused by the cornstalk borer (top image).

The History of Genetic Science

Though genetic editing is "cutting-edge" science, the process of influencing the genetic expression of organisms is not really a new concept. The idea of "inheritance," meaning that characteristics can be passed down and altered through reproduction, was discovered very early in history and as early as 10,000 years ago, humans were already engaging in genetic manipulation through selective breeding and propagation.

The global food crop known as "corn" provides an example of how these ancient experiments shaped human life. The common corn plant is classified as the species *Zea mays*, but scientists know that the species grown today bears little resemblance to the ancestor species that existed in North America before humans began growing corn as a food crop. Over the centuries, humans bred corn to have larger fruiting bodies with juicier, more flavorsome seeds. They did this by selecting individuals with larger fruits or that were more productive and harvesting the seeds from these selected individuals to propagate the next generation. Most of the species utilized in agriculture were produced in a similar way to corn, through selective propagation towards the goal of enhancing certain characteristics.

It was also ancient humans who created the first genetically modified (GM) animals by selecting individual animals to breed based on their traits. This is how humanity created cows, pigs, and horses, using lineages of animals bred to enhance their useful traits. Similarly, humanity created all the breeds and varieties of domestic dogs and cats utilizing this simple form of genetic engineering. In the twenty-first century, people are still utilizing the same techniques to create new animals, such as how tropical fish breeders select individuals or even mix species in an effort to create new "breeds" of fish.

Undoubtedly, manipulating genetics had tremendous influence on human life, shaping modern culture in many ways, but there are many ways in which genetic engineering has been problematic. The overpopulation of "pet" animals, hybrids and domesticated versions that cannot live in the wild, has created ecological problems around the world and many consider the breeding and harvesting of food animals as an immoral and unethical industry. Likewise, the cultivation of food plants facilitated environmental devastation in the form of monocultures and resulted in crops with reduced nutritional value.

Beginning in the late twentieth century, science and human agriculture merged through the implementation of new techniques that allowed humans to directly influence genes. This form of direct genetic modification also ushered in a new era of genetic manipulation targeting the human species and in which scientists seek to use genetic modification in medical treatment.

Discovering and Mastering the Gene

Genes are segments of genetic code stored within the cells of all organisms that contribute to certain characteristics displayed by the organism. The shape of a person's chin, the color of their hair, or even personality traits, are shaped, to some degree, by the "expression" of genes, which means the way that genes direct the development of bodies and minds.

The history of genetics can be traced to ancient Greece, where first imagined that there was some sort of atomic system in which parents were contributing some "component" to any resulting offspring. In the eighteenth century, work on plant and animal breeding led to the scientific documentation of phenomena that pointed the way towards the discovery of genes. It was Gregor Mendel who famously put these various pieces of evidence together in the 1850s and came up with the idea of genetic inheritance. Mendel introduced concepts such as "dominant" and "recessive" genes and helped to show how traits changed and were passed down through generations within a species or population.[1]

Then, in 1869, Swiss chemist Friedrich Miescher identified a substance inside white blood cells that would later become known as "nucleic acid," and, through further discovery, as "deoxyribonucleic acid" or "DNA." This, it was discovered, along with another substance called "ribonucleic acid" or "RNA" formed the basis of human inheritance and contained the "code," to use a simplified metaphor, for the expression of human physical properties.[2]

The discovery of genetics and of DNA and RNA changed the landscape of our understanding of human characteristics and bodies. Over the century that followed, humanity developed the ability to determine the sequences of individual genes and scientists got better and better and tracing certain genes or collections of genes to certain human traits. In the 1990s and early 2000s, many believed that there was a single, specific gene for many human traits. For many years, it was common to hear Americans speculate about a gene for blue eyes, or for baldness, or even for homosexuality. Many believed that when it was possible to find these genes, it would also become possible to change the traits connected to those genes. A person might therefore be able to genetically change their eye color to blue, or scientists might genetically cure baldness. On the darker side, prejudiced people speculated about using genetic manipulation to cure "homosexuality," which is neither a disorder that needs to be cured, nor a purely physical, genetic phenomenon. Over the years, scientists have found that genetic inheritance isn't as simple as was once believed and that it is rare for any trait to be controlled by just a single gene. The expression of physical traits like eye color and baldness are now understood to be connected to a number of different genes that collectively influence expression. More complex aspects of human existence, like personality traits or sexual preference, have genetic components but are not the matter of simple genetic expression alone.

However, while the idea of finding "the gene" for baldness and other things proved to be more illusory than productive, scientists were able to link many serious medical conditions to various genes or combinations of genes and further were able to discover how the absence of genes can affect the body. It was in the 1960s that scientists first speculated about the possibility of "injecting" or "inserting" gene sequences into the human body to change how the body behaves. The research paper "Gene Therapy for Human Genetic Disease?" was published in the popular journal *Science* in 1972, with scientists Theodore Friedmann and Richard Roblin presenting the basic theory of how this might work, but also warning that active development of gene therapy was not advised because of the lack of understanding about how genes worked.

Around eighteen years later, in 1990, a four-year-old girl, Ashanthi DeSilva, became the first patient treated with gene therapy. DeSilva suffered from a rare genetic disorder called "severe combined immunodeficiency," which essentially meant that she could barely spend time with others without risking severe illness or death. Scientists discovered that this disease stemmed from the lack of an enzyme adenosine deaminase (ASA), which was created by way of a certain gene. Using a virus as a "vector" or "carrier," scientists were able to insert a functional copy of this gene into DeSilva's body, which allowed her to produce ASA herself, and increased the function of her immune system.[3]

This 1990 treatment was the beginning of gene therapy, but the field stalled in 1999 after the death of Jesse Gelsinger, a patient at the University of Pennsylvania suffering from ornithine transcarbamylase deficiency. When Gelsinger died, from an immune system reaction, the FDA suspended the University of Pennsylvania's entire gene therapy operation, which was, at the time, one of the largest and most productive in the country. Researchers turned their attention to looking into the safety concerns surrounding genetic treatments and, specifically, the use of viral vectors and development largely stopped for a few years.[4]

In the mid-2000s, gene therapy research began to resume in many countries and the Food and Drug Administration (FDA) again began to approve genetic treatments for certain disorders. Part of what changed, in the ensuing years, was that viral vector technology improved and became more reliable. The development of the viral vectors known as adeno-associated viruses (AAVs) was a big part of the return of gene therapy, providing a more targeted delivery system. In addition, utilizing the eye to introduce viral vectors proved to be another advancement, as eyes are relatively limited systems that typically don't allow introduced materials to reach other areas as easily. Advancements like these led to an explosion of development in gene therapy around the world.

Much of the modern focus on gene therapy focuses on Clustered Regularly Interspaced Short Palindromic Repeats (CRISPR) a system that can be used to edit genes. In traditional gene therapy, a functional copy of a gene is introduced, but this does not eliminate dysfunctional gene activity. Using CRISPR, which is a "gene-editing" technique, it is possible to eliminate dysfunctional genes from a cell. This is therefore a different, though related way, to eliminate a genetic disorder, by eliminating the genetic abnormality at the root of the dysfunction. The CRISPR program gained worldwide attention in the late 2010s as scientists began working on gene-editing solutions involving sickle cell anemia.[5]

Editing Human Life

While gene therapy and genetic editing have provided promising results when it comes to combating genetic diseases and disorders, the capability to edit the genome has also raised ethical issues that Americans are still struggling to understand. Some believe that the genetic manipulation of living organisms is unnatural or violates national patterns and processes. Others believe that humanity is meddling with biological and physical properties about which humans know very little and that there may yet be unforeseen consequences.

Ethical concern over genetic manipulation techniques can be seen in the ongoing debate over the cloning of animals and the potential to clone humans, which many Americans and citizens around the world consider a potentially dangerous or immoral path of development. Cloned organisms have demonstrated a high likelihood of developing genetic disorders, which is one of the primary ethical objections against experimentation with this kind of technology. In addition, many simply consider cloning to be unnatural or contrary to either the perceived "laws of nature" or some kind of "spiritual" principle against the technological creation of life. A similar controversy has developed over projects designed to recreate extinct organisms by using genetic manipulation to essentially "clone" extinct species. While there are numerous scientists working on "de-extinction" technology, many consider this process potentially dangerous and unethical given that it is unclear how a recreated species would survive or whether this process can be conducted without producing genetic abnormalities that might amount to animal cruelty.

Another controversy surrounding genetic editing and manipulation techniques concerns that possibility of eliminating genetic abnormalities from fetuses. While in some cases members of the public support such efforts, such as in the effort to eliminate genetic dysfunctions that negatively impact a person's quality of life, many have expressed concern that this kind of technology will be used to eliminate aspects of human life that differ from the norm but cannot be plainly labeled as dysfunction. For instance, if gene editing could eliminate deafness, some might consider this a benefit, but members of the deaf community have objected to characterizing deafness as a disability and instead argue that deafness is part of the natural diversity of human life. This view of characteristics like deafness, autism, or blindness is part of a movement known as "neurodiversity," which promotes the idea that human minds and bodies can work in various different ways without being considered dysfunctional or problematic in a way that needs to be "cured."

A Historical Stigma

The potential negative impact of genetic manipulation seen from past efforts colors the way that Americans and people around the world view genetic science in the twenty-first century as well. Many critics worry that unchecked genetic experimentation will lead to the unintentional increase in genetic abnormalities or illness. Others are concerned that genetic manipulation violates natural or spiritual principles. Others are concerned that this technology could fall into the "wrong hands" becoming a weapon. The historical development of genetic manipulation techniques demonstrates that the technology has long been controversial and potentially problematic, but also demonstrates that genetic medicine and development might also hold the key to addressing some of humanity's most ancient threats.

Works Used

Gostimskaya, Irina. "CRISPR-Cas9: A History of Its Discovery and Ethical Considerations of Its Use in Genome Editing." *Biochemistry* 87, no. 8, Aug. 15, 2022. www.ncbi.nlm.nih.gov/pmc/articles/PMC9377665/. Accessed Aug. 26, 2023.

Miko, Ilona. "Gregor Mendel and the Principles of Inheritance." *Nature*. 2008. www.nature.com/scitable/topicpage/gregor-mendel-and-the-principles-of-inheritance-593/. Accessed Aug. 25, 2023.

Mitha, Farhan. "The Return of Gene Therapy." *Labiotech*. Nov. 4, 2020. www.labiotech.eu/in-depth/gene-therapy-history/. Accessed Aug. 25, 2023.

Pray, Leslie A. "Discovery of DNA Structure and Function: Watson and Crick." *Nature*. 2008. www.nature.com/scitable/topicpage/discovery-of-dna-structure-and-function-watson-397/. Accessed Aug. 25, 2023.

Rinde, Meir. "The Death of Jesse Gelsinger, 20 Years Later." *Science History Institute*. June 4, 2019. sciencehistory.org/stories/magazine/the-death-of-jesse-gelsinger-20-years-later/. Accessed Aug. 26, 2023.

Notes

1. Miko, "Gregor Mendel and the Principles of Inheritance."
2. Pray, "Discovery of DNA Structure and Function: Watson and Crick."
3. Mitha, "The Return of Gene Therapy."
4. Rinde, "The Death of Jesse Gilsinger, 20 Years Later."
5. Gostimskaya, "CRISPR-Cas9: A History of Its Discovery and Ethical Consideration of Its Use in Genome Editing."

What Is Genome Editing?

National Human Genome Research Institute/NIH, 2021

Genome editing is a method that lets scientists change the DNA of many organisms, including plants, bacteria, and animals. Editing DNA can lead to changes in physical traits, like eye color, and disease risk. Scientists use different technologies to do this.

Overview

Genome editing technologies enable scientists to make changes to DNA, leading to changes in physical traits, like eye color, and disease risk. Scientists use different technologies to do this. These technologies act like scissors, cutting the DNA at a specific spot. Then scientists can remove, add, or replace the DNA where it was cut.

The first genome editing technologies were developed in the late 1900s. More recently, a new genome editing tool called CRISPR, invented in 2009, has made it easier than ever to edit DNA. CRISPR is simpler, faster, cheaper, and more accurate than older genome editing methods. Many scientists who perform genome editing now use CRISPR.

In the Laboratory

One way that scientists use genome editing is to investigate different diseases that affect humans. They edit the genomes of animals, like mice and zebrafish, because animals have many of the same genes as humans. For example, mice and humans share about 85 percent of their genes! By changing a single gene or multiple genes in a mouse, scientists can observe how these changes affect the mouse's health and predict how similar changes in human genomes might affect human health.

Scientists at the National Human Genome Research Institute (NHGRI) are doing just this. The Burgess lab, for example, is studying zebrafish genomes. Scientists in this lab delete different genes in zebrafish one at a time using CRISPR to see how the deletion impacts the fish. The Burgess lab focuses on 50 zebrafish genes which are similar to the genes that cause human deafness so that they can better understand the genomic basis of deafness.

Treating Disease

Scientists are developing gene therapies—treatments involving genome editing—to prevent and treat diseases in humans. Genome editing tools have the potential to help treat diseases with a genomic basis, like cystic fibrosis and diabetes. There

> What if gene therapies are too expensive and only wealthy people can access and afford them?

are two different categories of gene therapies: germline therapy and somatic therapy. Germline therapies change DNA in reproductive cells (like sperm and eggs). Changes to the DNA of reproductive cells are passed down from generation to generation. Somatic therapies, on the other hand, target non-reproductive cells, and changes made in these cells affect only the person who receives the gene therapy.

In 2015, scientists successfully used somatic gene therapy when a one-year old in the United Kingdom named Layla received a gene editing treatment to help her fight leukemia, a type of cancer. These scientists did not use CRISPR to treat Layla, and instead used another genome editing technology called TALENs. Doctors tried many treatments before this, but none of them seemed to work, so scientists received special permission to treat Layla using gene therapy. This therapy saved Layla›s life. However, treatments like the one that Layla received are still experimental because the scientific community and policymakers still have to address technical barriers and ethical concerns surrounding genome editing.

Technical Barriers

Even though CRISPR improved upon older genome editing technologies, it is not perfect. For example, sometimes genome editing tools cut in the wrong spot. Scientists are not yet sure how these errors might affect patients. Assessing the safety of gene therapies and improving upon genome editing technologies are critical steps to ensure that this technology is ready for use in patients.

Ethical Concerns

Scientists and all of us should carefully consider the many ethical concerns that can emerge with genome editing, including safety. First and foremost, genome editing must be safe before it is used to treat patients. Some other ethical questions that scientists and society must consider are:

1. Is it okay to use gene therapy on an embryo when it is impossible to get permission from the embryo for treatment? Is getting permission from the parents enough?

2. What if gene therapies are too expensive and only wealthy people can access and afford them? That could worsen existing health inequalities between the rich and poor.

3. Will some people use genome editing for traits not important for health, such as athletic ability or height? Is that okay?

4. Should scientists ever be able to edit germline cells? Edits in the germline would be passed down through generations.

Most people agree that scientists should not edit the genomes of germline cells at this time because the safety and Scientific communities across the world are approaching germline therapy research with caution because edits to a germline cell would be passed down through generations. Many countries and organizations have strict regulations to prevent germline editing for this reason. The NIH, for example, does not fund research to edit human embryos.

Print Citations

CMS: National Human Genome Research Institute/NIH. "What Is Genome Editing?" In *The Reference Shelf: Gene Editing & Genetic Engineering*, edited by Micah L. Issitt, 9–11. Amenia, NY: Grey House Publishing, 2023.

MLA: National Human Genome Research Institute/NIH. "What Is Genome Editing?" *The Reference Shelf: Gene Editing & Genetic Engineering*, edited by Micah L. Issitt, Grey House Publishing, 2023, pp. 9–11.

APA: National Human Genome Research Institute/NIH. (2023). What is genome editing? In Micah L. Issitt (Ed.), *The reference shelf: Gene editing & genetic engineering* (pp. 9–11). Amenia, NY: Grey House Publishing. (Original work published 2021)

A Brief History of Human Gene Editing

By Carmen Leitch
Labroots, July 31, 2022

DNA was characterized in 1953 by James Watson, Francis Crick, Maurice Wilkins, and Rosalind Franklin at King's College London. The nature of the genetic code, which uses codons, strings of three nucleotide bases, to encode for an amino acid and generate protein was discovered by Marshall Nirenberg, Har Khorana, and Severo Ochoa at the National Institutes of Health (NIH).

There is variation and redundancy in the genetic code, so it is possible for benign genetic changes to occur. But changes in the sequence can also cause serious disease, usually when those changes, or mutations, result in alterations in a protein that a gene encodes for. Nonfunctional or dysfunctional proteins can have huge consequences. Scientists have traced many different diseases back to specific errors in the genetic code. If those errors could be fixed, or functional copies of a mutated gene could be delivered to cells, it may be possible to cure such diseases. This is the basic idea behind gene therapy.

Martin Cline of the University of California, Los Angeles was the first person to transfer a functional gene into mice to create a transgenic organism. Cline was also the first person to test genetic engineering on humans, against NIH guidelines. In his approach, the beta-globin gene was transferred into the cells of two patients with beta-thalassamia. The experiment failed in part because the treated cells did not replicate. This incident led the NIH to warn other researchers to not attempt similar experiments. Cline lost grants and resigned a department chairmanship at UCLA.

Scientists realized that if gene therapy was going to work, something had to deliver gene therapy reagents into the nucleus of cells, where the genome is. Viral vectors were created for this purpose.

The first time a gene was successfully transferred into the nucleus of human cells was in May 1989. Clinicians inserted a neomycin-resistance gene to mark therapeutic cells that were then infused into a small group of patients. These cells remained in circulation for about two months and caused no adverse reactions.

In 1990, two patients with forms of severe combined immunodeficiency disorder (sometimes known as bubble-boy disease) were selected to be part of a trial. Four-year-old Ashanti DeSilva was first treated on September 14, 1990, and another patient was treated about nine months later.

In this approach, T cells were removed from the patient, and treated with genetic engineering to correct the mutation in the DNA of those T cells. The cells were then reinserted into the patient. For about two years, DeSilva received 11 infusions of the treatment. The second patient got 12 infusions over 18 months. Unfortunately, the engineered T cells would never replicate. But while the patients were not cured, T cells counts improved, and both survived to adulthood. Researchers learned more about how to create a successful technique. DeSilva is now a genetic counselor.

> **Scientists realized that if gene therapy was going to work, something had to deliver gene therapy reagents into the nucleus of cells, where the genome is—viral vectors were created for this purpose.**

New viral vectors were created and tested in another gene therapy trial in 2000. This time, there would be tragedy when four of the nine patients developed leukemia. Scientists quickly learned that any viral vectors that would be used in humans had to be carefully and thoroughly tested first.

China approved Gendicine, its first gene therapy in 2003, for certain forms of skin cancer. A protein that can fight tumors is carried by a viral vector to tumor cells. There, the gene therapy is meant to stimulate the activity of genes that suppress tumors and influence the immune response.

In 2009, scientists led by Jean Bennett of the University of Pennsylvania treated patients with a genetic eye disorder using an AAV. This trial would lead, eight years later, to the first FDA-approved gene therapy. The first in vivo gene addition therapy, Luxturna, was approved by the FDA in 2017, to treat patients with a form of inherited blindness called biallelic RPE65 mutation-associated retinal dystrophy. The normal version of RPE65 is delivered to cells in this treatment, so they can express a protein that is required for the conversion of light to electrical signals, enabling vision.

On August 30, 2017, the FDA approved Kymriah (tisagenlecleucel) as a treatment for some young patients with a type of acute lymphoblastic leukemia (ALL). In this case, a new gene is introduced into T cells so they will be better at targeting cancer.

Scientists have continued to work to develop even more methods, including CRISPR, to open up more treatment options and experimental gene therapies for patients.

Print Citations

CMS: Leitch, Carmen. "A Brief History of Human Gene Editing." In *The Reference Shelf: Gene Editing & Genetic Engineering,* edited by Micah L. Issitt, 12–14. Amenia, NY: Grey House Publishing, 2023.

MLA: Leitch, Carmen. "A Brief History of Human Gene Editing." *The Reference Shelf: Gene Editing & Genetic Engineering,* edited by Micah L. Issitt, Grey House Publishing, 2023, pp. 12–14.

APA: Leitch, C. (2023). A brief history of human gene editing. In Micah L. Issitt (Ed.), *The reference shelf: Gene editing & genetic engineering* (pp. 12–14). Amenia, NY: Grey House Publishing. (Original work published 2022)

Science and History of GMOs and Other Food Modification Processes

FDA, April 19, 2023

Did You Know?

Genetic engineering is often used in combination with traditional breeding to produce the genetically engineered plant varieties on the market today.

How Has Genetic Engineering Changed Plant and Animal Breeding?

For thousands of years, humans have been using traditional modification methods like selective breeding and cross-breeding to breed plants and animals with more desirable traits. For example, early farmers developed cross-breeding methods to grow corn with a range of colors, sizes, and uses. Today's strawberries are a cross between a strawberry species native to North America and a strawberry species native to South America.

Most of the foods we eat today were created through traditional breeding methods. But changing plants and animals through traditional breeding can take a long time, and it is difficult to make very specific changes. After scientists developed genetic engineering in the 1970s, they were able to make similar changes in a more specific way and in a shorter amount of time.

A Timeline of Genetic Modification in Agriculture

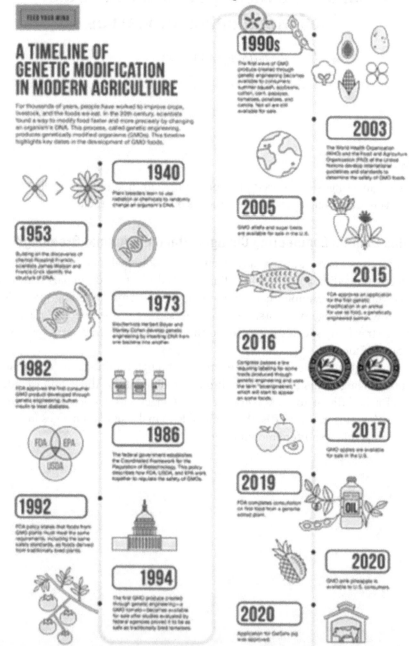

Circa 8000 BCE: Humans use traditional modification methods like selective breeding and cross-breeding to breed plants and animals with more desirable traits.

1866: Gregor Mendel, an Austrian monk, breeds two different types of peas and identifies the basic process of genetics.

1922: The first hybrid corn is produced and sold commercially.

1940: Plant breeders learn to use radiation or chemicals to randomly change an organism's DNA.

1953: Building on the discoveries of chemist Rosalind Franklin, scientists James Watson and Francis Crick identify the structure of DNA.

> Genetic engineering is a process that involves identifying the genetic information, or "gene"; copying that information from the organism that has the trait; inserting that information in the DNA of another organism; and then growing the new organism.

1973: Biochemists Herbert Boyer and Stanley Cohen develop genetic engineering by inserting DNA from one bacteria into another.

1982: FDA approves the first consumer genetically modified organism (GMO) product developed through genetic engineering: human insulin to treat diabetes.

1986: The federal government establishes the Coordinated Framework for the Regulation of Biotechnology. This policy describes how the U.S. Food and Drug Administration (FDA), U.S. Environmental Protection Agency (EPA), and U.S. Department of Agriculture (USDA) work together to regulate the safety of GMOs.

1992: FDA policy states that foods from GMO plants must meet the same requirements, including the same safety standards, as foods derived from traditionally bred plants.

1994: The first GMO produce created through genetic engineering—a GMO tomato—becomes available for sale after studies evaluated by federal agencies proved it to be as safe as traditionally bred tomatoes.

1990s: The first wave of GMO produce created through genetic engineering becomes available to consumers: summer squash, soybeans, cotton, corn, papayas, tomatoes, potatoes, and canola. Not all are still available for sale.

2003: The World Health Organization (WHO) and the Food and Agriculture Organization (FAO) of the United Nations develop international guidelines and standards to determine the safety of GMO foods.

2005: GMO alfalfa and sugar beets are available for sale in the United States.

2015: FDA approves an application for the first genetic modification in an animal for use as food, a genetically engineered salmon.

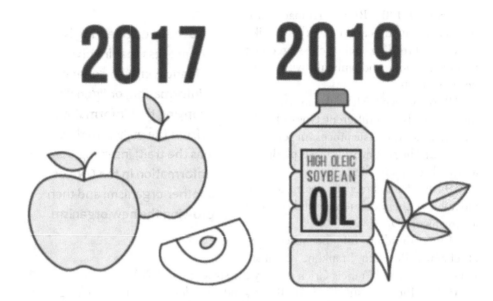

2016: Congress passes a law requiring labeling for some foods produced through genetic engineering and uses the term "bioengineered," which will start to appear on some foods.

2017: GMO apples are available for sale in the U.S.

2019: FDA completes consultation on first food from a genome edited plant.

2020: GMO pink pineapple is available to U.S. consumers.

2020: Application for GalSafe pig was approved.

How Are GMOs Made?

"GMO" (genetically modified organism) has become the common term consumers and popular media use to describe foods that have been created through genetic engineering. Genetic engineering is a process that involves:

- Identifying the genetic information—or "gene"—that gives an organism (plant, animal, or microorganism) a desired trait

- Copying that information from the organism that has the trait

- Inserting that information into the DNA of another organism

- Then growing the new organism

Making a GMO Plant, Step by Step

The following example gives a general idea of the steps it takes to create a GMO plant. This example uses a type of insect-resistant corn called "Bt corn." Keep in mind that the processes for creating a GMO plant, animal, or microorganism may be different.

Identify

To produce a GMO plant, scientists first identify what trait they want that plant to have, such as resistance to drought, herbicides, or insects. Then, they find an organism (plant, animal, or microorganism) that already has that trait within its genes. In this example, scientists wanted to create insect-resistant corn to reduce the need to spray pesticides. They identified a gene in a soil bacterium called Bacillus thuringiensis (Bt), which produces a natural insecticide that has been in use for many years in traditional and organic agriculture.

Copy

After scientists find the gene with the desired trait, they copy that gene.

For Bt corn, they copied the gene in Bt that would provide the insect-resistance trait.

Insert

Next, scientists use tools to insert the gene into the DNA of the plant. By inserting the Bt gene into the DNA of the corn plant, scientists gave it the insect resistance trait.

This new trait does not change the other existing traits.

Grow

In the laboratory, scientists grow the new corn plant to ensure it has adopted the desired trait (insect resistance). If successful, scientists first grow and monitor the new corn plant (now called Bt corn because it contains a gene from Bacillus thuringiensis) in greenhouses and then in small field tests before moving it into larger field tests. GMO plants go through in-depth review and tests before they are ready to be sold to farmers.

The entire process of bringing a GMO plant to the marketplace takes several years.

What Are the Latest Scientific Advances in Plant and Animal Breeding?

Scientists are developing new ways to create new varieties of crops and animals using a process called genome editing. These techniques can make changes more quickly and precisely than traditional breeding methods.

There are several genome editing tools, such as CRISPR. Scientists can use these newer genome editing tools to make crops more nutritious, drought tolerant, and resistant to insect pests and diseases.

Print Citations

CMS: FDA. "Science and History of GMOs and Other Food Modification Processes." In *The Reference Shelf: Gene Editing & Genetic Engineering*, edited by Micah L. Issitt, 15–21. Amenia, NY: Grey House Publishing, 2023.

MLA: FDA. "Science and History of GMOs and Other Food Modification Processes." *The Reference Shelf: Gene Editing & Genetic Engineering*, edited by Micah L. Issitt, Grey House Publishing, 2023, pp. 15–21.

APA: FDA. (2023). Science and history of GMOs and other food modification processes. In Micah L. Issitt (Ed.), *The reference shelf: Gene editing & genetic engineering* (pp. 15–21). Amenia, NY: Grey House Publishing. (Original work published 2023)

Did He Jiankui "Make People Better"? Documentary Spurs a New Look at the Case of the First Gene-Edited Babies

By G. Owen Schaefer
The Conversation, December 20, 2022

In the four years since an experiment by disgraced scientist He Jiankui resulted in the birth of the first babies with edited genes, numerous articles, books and international commissions have reflected on whether and how heritable genome editing—that is, modifying genes that will be passed on to the next generation—should proceed. They've reinforced an international consensus that it's premature to proceed with heritable genome editing. Yet, concern remains that some individuals might buck that consensus and recklessly forge ahead—just as He Jiankui did.

Some observers—myself included—have characterized He as a rogue. However, the new documentary "Make People Better," directed by filmmaker Cody Sheehy, leans toward a different narrative. In its telling, He was a misguided centerpiece of a broader ecosystem that subtly and implicitly supported rapid advancement in gene editing and reproductive technologies. That same system threw He under the bus—and into prison—when it became evident that the global community strongly rejected his experiments.

Creation of the "CRISPR Babies"

"Make People Better" outlines an already well-documented saga, tracing the path of He from a promising young scientist at Rice and Stanford to a driven researcher establishing a laboratory in China that secretly worked to make heritable genome editing a reality.

He's experiment involved using the CRISPR-Cas9 technique. Sometimes compared to "molecular scissors," this precision tool allows scientists to make very specific edits to DNA in living cells. He used CRISPR to alter the CCR5 gene in human embryos with the goal of conferring immunity to HIV. These embryos were brought to term, resulting in the birth of at least three children with altered DNA.

The revelation of the births of the first gene-edited babies in November 2018 resulted in an international uproar. A laundry list of ethical failings in He's experiment quickly became evident. There was insufficient proof that editing embryos with CRISPR was safe enough to be done in humans. Appropriate regulatory

---

approval had not been obtained. The parents' consent was grossly inadequate. And the whole endeavor was shrouded in secrecy.

New Context, Same Story

Three figures play a central role in "Make People Better"'s study of He Jiankui. There's Antonio Regalado, the reporter from MIT Technology Review who broke the original story. There's Ben Hurlbut, an ethicist and confidante of He. And there's Ryan (the documentary withholds his full identity), a public relations representative who worked with He to make gene editing palatable to the world. He Jiankui himself was not interviewed, though his voice permeates the documentary in previously unreleased recordings by Hurlbut.

Regalado and Hurlbut have already written a considerable amount on this saga, so the documentary's most novel contribution comes from Ryan's discussion of his public relations work with He. Ryan appears to be a true believer in He's vision to literally "make people better" by using gene editing to prevent dreadful diseases.

But Ryan is aware that public backlash could torpedo this promising work. His reference point is the initial public hostility to GMO foods, and Ryan strove to avoid that outcome by gradually easing the public in to the heritable gene editing experiment.

This strategy turned out to be badly mistaken for a variety of reasons. He Jiankui was himself eager to publicize his work. Meanwhile, Regalado's tenacious journalism led him to a clinical trials registry where He had quietly posted about the study.

But ultimately, those factors just affected the timing of revelation. Both Ryan and He failed to appreciate that they had very little ability to influence how the experiment would be received, nor how much condemnation would result.

Blind Spots

While some documentaries strive to be flies on the wall, objectivity is elusive. Tone, framing, editing and choice of interview subjects all coalesce into a narrative with a perspective on the subject matter. A point of view is not itself objectionable, but it opens the documentary to critiques of its implicit stance.

An uncomfortable tension lies at the center of "Make People Better."

On the one hand, the documentary gives substantial attention to Hurlbut and Ryan, who emphasize that He did not act alone. He discussed his plans with dozens of people in China and around the world, whose implicit support was essential to both the experiment and his confidence that he was doing nothing wrong.

On the other hand, the documentary focuses on understanding He's background, motives and ultimate fate. Other figures who might have influenced He to take a different path fade into the background—sometimes quite literally, appearing for only seconds before the documentary moves on.

Indeed, as a biomedical ethicist, I believe there is good reason to put responsibility for the debacle squarely on He's shoulders. Before the news broke in 2018, international panels of experts had already issued advisory statements that heritable

gene editing was premature. Individuals like Hurlbut personally advised He as
much. The secrecy of the experiment itself
is a testament: He must have suspected
the international community would reject
the experiment if they knew what was going on.

> **Gene editing could save lives by reducing the incidence of certain diseases.**

If He had gone through proper, transparent channels—preregistering the trial and consulting publicly with international experts on his plans before he began—
the whole saga could have been averted. He chose a different, more dangerous
and secretive path from the vast majority of researchers working in reproductive
biotechnology, which I suggest must be acknowledged.

The documentary does not reflect critically on its own title. The origin of the
phrase "make people better" is surprising and the film's most clever narrative moment, so I won't spoil it. But does heritable gene editing really make people better? Perhaps instead, it makes better people.

The gene-edited babies were created via in vitro fertilization specifically as a part
of He's experiment. They would not have existed if He had never gotten involved in
gene editing. So, some would argue, He did not save any individual from contracting
HIV. Rather, he created new people potentially less likely to contract HIV than the
general population.

I contend that this doesn't mean gene editing is pointless. From a population
health perspective, gene editing could save lives by reducing the incidence of certain diseases. But this perspective does change the moral tenor of gene editing,
perhaps reducing its urgency.

What's more, editing CCR5 is a dubious means to improve human well-being,
since there are already effective ways to prevent HIV infection that are far less risky
and uncertain than heritable gene editing. Scientific consensus suggests that the
best first-in-human candidates for heritable gene editing are instead devastating
genetic disorders that cannot be ameliorated in other ways.

The Future for He Jiankui

Perhaps due to the timing of its filming, the documentary does not dwell on He
being sentenced to three years in Chinese prison as a result of the experiment, nor
mention that he was released early in 2022.

Evidently, He is not content to fade quietly into obscurity. He says he is slated in
March 2023 to give a talk at the University of Oxford that may shed more light on
his motives and actions. In the meantime, he has established a new biotech start-
up focused on developing gene therapies. To be clear, this work does not involve
editing embryos.

Still, it appears prison has not diminished He's ambition. He claims that he
could develop a cure for the degenerative genetic disease Duchenne muscular dystrophy—if he receives funding in excess of US$100 million.

To me, this ambition reflects a curious symmetry between Regalado and He in "Make People Better." Both are driven to be first, to be at the forefront of their respective fields. Sometimes, as with Regalado, this initiative can be good—his intrepid reporting and instinct to publish quickly brought He's unethical experiment to a rapid close. But in other cases, like He's, that drive can lead to dangerous science that runs roughshod over ethics and good governance.

Perhaps, then, the best lesson a viewer can take from "Make People Better" is that ambition is a double-edged sword. In the years to come, it will be up to the international community to keep such ambition in check and ensure proper restrictions and oversight on heritable genome editing.

Print Citations

CMS: Schaefer, G. Owen. "Did He Jiankui 'Make People Better'? Documentary Spurs a New Look at the Case of the First Gene-Edited Babies" In *The Reference Shelf: Gene Editing & Genetic Engineering,* edited by Micah L. Issitt, 22–25. Amenia, NY: Grey House Publishing, 2023.

MLA: Schaefer, G. Owen. "Did He Jiankui 'Make People Better'? Documentary Spurs a New Look at the Case of the First Gene-Edited Babies" *The Reference Shelf: Gene Editing & Genetic Engineering,* edited by Micah L. Issitt, Grey House Publishing, 2023, pp. 22–25.

APA: Schaefer, G. O. (2023). Did He Jiankui "make people better"? Documentary spurs a new look at the case of the first gene-edited babies. In Micah L. Issitt (Ed.), *The reference shelf: Gene editing & genetic engineering* (pp. 22–25). Amenia, NY: Grey House Publishing. (Original work published 2022)

Experts Weigh Medical Advances in Gene-Editing with Ethical Dilemmas

By Rob Stein
NPR, March 6, 2023

Hundreds of scientists, doctors, bioethicists, patients, and others started gathering in London Monday for the Third International Summit on Human Genome Editing. The summit this week will debate and possibly issue recommendations about the thorny issues raised by powerful new gene-editing technologies.

The last time the world's scientists gathered to debate the pros and cons of gene-editing—in Hong Kong in late 2018—He Jiankui, a biophysicist and researcher at Southern University of Science and Technology in Shenzhen, China, shocked his audience with a bombshell announcement. He had created the first gene-edited babies, he told the crowd—twin girls born from embryos he had modified using the gene-editing technique CRISPR.

He, who had trained at Rice University and Stanford, said he did it in hopes of protecting the girls from getting infected with the virus that causes AIDS. (The girls' father was HIV-positive.) But his announcement was immediately condemned as irresponsible human experimentation. Far too little research had been done, critics said, to know if altering the genetics of embryos in this way was safe. He ultimately was sentenced by a Chinese court to three years in prison for violating medical regulations.

In the more than four years since He's stunning announcement, scientists have continued to hone their gene-editing powers.

"A lot has happened over the last five years. It's been a busy period," says Robin Lovell-Badge from the Francis Crick Institute in London, who led the committee convening the new summit.

Doctors have made advances using CRISPR to try to treat or better understand many diseases, including devastating disorders like sickle cell disease, and conditions like heart disease and cancer that are even more common and influenced by genetics.

In recent years, scientists have produced new evidence about the risks and possible shortcomings of gene-editing, while also developing more sophisticated techniques that could be safer and more precise.

"We're at an exciting moment for sure with genome-editing," says Jennifer Doudna at the University of California, Berkeley, who helped discover CRISPR. "At the same time, we certainly have challenges."

"We Could Help a Lot of People"

One big remaining challenge and ethical question is whether scientists should ever again try to make gene-edited babies by modifying the DNA in human sperm, eggs or embryos. Such techniques, if successful could help families that have been plagued by devastating genetic disorders.

"There are more than 10,000 single genetic mutations that collectively affect probably hundreds of millions of people around the world," says Shoukhrat Mitalipov, a biologist at the Oregon Health and Science University in Portland who's been trying to find ways to safely gene-edit human embryos. "We could help a lot of people."

But the fear is a mistake could create new genetic diseases that could then be passed down for generations. Some scientists are also concerned about opening a slippery slope to "designer babies"—children whose parents try to pick and choose their traits.

"If we were to allow parents to genetically modify their children, we would be creating new groups of people who are different from each other biologically and some would have been modified in ways that are supposed to enhance them," says Marcy Darnovsky heads the Center for Genetics and Society in San Francisco. "And they would be—unfortunately I think—considered an enhanced race—a better group of people. And I think that could really just super-charge the inequities we already have in our world."

The Debate Among Mant Scientists Seems to Have Shifted to How to Edit a Genome Safely

Despite those concerns, some critics say the debate over the last five years has shifted from whether a prohibition on inheritable genetic modifications should ever be lifted to what technical hurdles need to be overcome to do it safely—and which diseases doctors might try to eradicate.

As evidence of that, the critics point to the fact that the subject of genetically modifying embryos, sperm or eggs to engineer modifications that would then be passed along to every subsequent generation is the focus of only one of three days of this summit—the first such conference since the CRISPR babies were announced.

"This is quite an ironic outcome," says Sheila Jasanoff is a professor of science and technology studies at Harvard's Kennedy School of Government.

"Instead of rejuvenating the calls to say: 'We should be much more careful,'" Jasanoff says, "it was as if the whole scientific community heaved a kind of sigh of relief and said: 'Well, look, of course there are limits. This guy has transgressed the limits. He's clearly outside the limits. And therefore everything else is now open for grabs. And therefore the problem before us now is to make sure that we lay out the guidelines and the rules.'"

Ben Hurlbut, a bioethicist at Arizona State University, agrees. "There was a time when this was considered taboo," he says. "But since the last summit, there's been a shift from asking the question of 'whether' to asking the question of 'how.'"

It Was Too Easy to Scapegoat He, Some Ethicists Say

Hurlbut and others also say scientists have failed to fully come to terms with the high-pressure environment of biomedical research that they say encouraged He to do what he did.

> The fear is a mistake could create new genetic diseases that could then be passed down for generations.

"It just feels easier to condemn He and say all bad resides in his person and he should be ostracized forever as we proceed apace. Not reckoning with what happened and why fosters a certain thoughtlessness, and I would say recklessness," Hurlbut says.

That lack of reckoning with what happened could be dangerous, critics say. It could, they fear, encourage others to try make more gene-edited babies, at a time when the public may never have been more skeptical about scientific experts.

"We have seen in recent years a sense that the experts have taken on too big a role and that they have tried to run roughshod over our day-to day-lives," says Hank Greely, a longtime Stanford University bioethicist. But whether or not inheritable genetic modifications should be allowed is "ultimately a decision for societies and not a decision for science."

A New Lab in Beijing

Meanwhile, He Jiankui appears to be trying to rehabilitate himself after serving his three-year prison sentence. He's set up a new lab in Beijing, is promising to develop new gene-therapies for diseases like muscular dystrophy, is giving scientific presentations, and is trying to raise money.

He's not expected to join the London summit this week, and is no longer talking about creating more gene-edited babies. Still, his activities are raising alarm in the scientific and bioethics communities. He declined NPR's request for an interview. But in a recently published interview with *The Guardian* the only regret he mentioned was in moving too fast.

"I'm concerned," Lovell-Badge says. "I'm surprised that that he's being allowed to practice science again. It just scares me."

Others agree.

"What he did was atrocious," says Dr. Kiran Musunuru, a professor of medicine at the University of Pennsylvania. "He shouldn't be allowed anywhere near a patient again. He's proven himself to be utterly unqualified."

Lovell-Badge and other organizers of the summit dispute criticisms that scientists are assuming gene-edited babies are inevitable and that the agenda for this week's conference short-changes a debate about the ethical and societal landmines that remain in this field of study.

Summit leaders say they'll dedicate the last day of the meeting to genetic modifications that can be passed down through generations; panel participants will feature scientists as well as a broad array of watchdog groups, patient advocates, bioethicists, sociologists and others.

Conference organizers say they have good reasons for focusing the first two-thirds of the meeting on the use of gene-editing to treat people who have already been born.

"The summit is a chance to really hear about what's happening in the field that has the greatest potential for improving human health," says R. Alta Charo, a professor emerta of law and bioethics from the University of Wisconsin, who helped organize the summit.

Questions of Equity Have Moved to Center Stage

But those current treatments raise their own ethical concerns—including questions of equity. Will the current and coming gene therapies be widely available, given how expensive and technologically complicated they can be to create and administer?

"We're not moving away from the conversation around heritable genome editing, but we are trying to shift some of that focus," says Francoise Baylis, a bioethicist who recently retired from Dalhousie University in Canada and helped plan the meeting. "Really important in this context is the issue of cost, because we have been seeing gene-therapies come onto the market with million-dollar price tags. That's not going to be available to the average person."

The availability of gene-therapy treatments in lower-income countries must be a focus of concern, Baylis says.

"We're going to be asking questions about where are the people who are most likely to be benefit," she says, "and are they going to have access?"

Print Citations

CMS: Stein, Rob. "Experts Weigh Medical Advances in Gene-Editing with Ethical Dilemmas." In *The Reference Shelf: Gene Editing & Genetic Engineering,* edited by Micah L. Issitt, 26–29. Amenia, NY: Grey House Publishing, 2023.

MLA Stein, Rob. "Experts Weigh Medical Advances in Gene-Editing with Ethical Dilemmas." *The Reference Shelf: Gene Editing & Genetic Engineering,* edited by Micah L. Issitt, Grey House Publishing, 2023, pp. 26–29.

APA: Stein, R. (2023). Experts weigh medical advances in gene-editing with ethical dilemmas. In Micah L. Issitt (Ed.), *The reference shelf Gene editing & genetic engineering* (pp. 26–29). Amenia, NY: Grey House Publishing. (Original work published 2023)

2
Pioneering Science

Above, jellyfish at the Wild Life Sydney Zoo. Photo by Ank Kumar, CC BY-SA 4.0, via Wikimedia.

The applications of bioluminescent bacteria include biosensors for the detection of contaminants, measurement of pollutant toxicity, and the monitoring of genetically engineered bacteria released into the environment.

New Developments in Genetic Technology

While genetic modification is an ancient practice, the discovery of genes and the development of modern "gene-editing" technologies, like Clustered Regularly Interspaced Short Palindromic Repeats (CRISPR), have pushed the science of genetic modification in new directions and has made practical genetic medicine a reality. From 1990 to the 2020s, genetic medicine evolved from a futuristic possibility to one of the biggest areas of technological development, but as of the 2020s, this has not meant that genetic medicine was available for everyone.

In addition, while genetically modified organisms (GMOs) have been part of the human agricultural industry for thousands of years, new technologies are emerging to help protect agriculture and the food supply from changing environmental realities, including the climate change crisis. In addition, biotechnology is not producing products that could impact other arenas of human industry, like textiles and manufacturing, as well as environmental bioengineered products that might help humanity to manage the current climate and pollution crisis.

The Practicality of Genetic Medicine

Genetic medicine has tremendous promise, but as of yet, genetic treatments have only begun to influence modern medicine. While great strides have been made in developing genetic testing and treatment, these developments have had little impact on American medicine in general, largely due to lack of access. Barriers at the provider and economic levels have thus far prevented genetic treatments from becoming mainstream and addressing these barriers is considered, by experts, to be one of the primary challenges in the field.

At the provider level, many medical offices and institutions lack access to genetic treatment options or lack specialists to provide genetic treatment to patients. The lack of specialists and training in advanced genetic medical techniques contributes to the lack of access for many providers and hospitals. Studies have also found a broad lack of personnel working in genetic services, at various levels. Medical schools have vacancies in medical genetics programs and many of those who train in genetic medicine pursue careers in biotechnology, which offers higher pay, rather than utilizing their training to help bring genetic medicine to patients. This is true also for genetic testing with a national shortage of staff at companies and institutions capable of interpreting genetic data obtained through testing. Overall, this means that patients suffering from advanced genetic disorders may be unable to access genetic testing and specialized genetic care or may experience long wait times when trying to access genetic medical treatments or services.[1]

Lack of access is much more prominent in underserved communities and, like many arenas of modern medicine, those with economic advantages are far better able to access genetic medical treatments. In addition, specialists with training in genetic services are more likely to gravitate towards higher-income institutions or to work in the private sector in technological development and this reduces the number of specialists available to work directly with patients in middle- and low-income areas or communities. Further, when genetic treatments become available, many insurance companies are slow to recognize and provide coverage for genetic options. In some cases, even genetic testing may not be included in some insurance plans or may still require significant payments from patients.[2]

While many barriers remain to making genetic medicine a more important part of standard American primary care, on the research side of the equation, the technology to create and utilize altered genes or edited genes has continued to improve. Viral vectors—virus segments utilized to introduce genes to organisms—remain one of the most fruitful areas of research in the field but questions remain about the practicality and efficacy of modern systems. As of 2023, there are more than 100 gene-therapy agents in testing or in various stages of development, which expresses the high level of potential that biotechnologists see in this field.

In addition, the CRISPR gene-editing system, which has become the major focus in many different areas of genetic medicine, is also a technology very much in its infancy. In the 2020s, scientists are looking for ways to increase the precision and effectiveness of CRISPR-introduced genes and this is one of the most promising and active areas of study within the genetic development discipline.[3]

Expanding into New Areas

While the potential of genetic medicine has generated the most excitement, there are many other ways in which genetic editing and modification could play a role in addressing major issues facing humanity. Environmental bioengineering is an emerging field that applies the principles of genetic modification and engineering to address environmental problems.[4]

As of 2023, scientists and biotech companies are looking into genetic modification to address pollution and waste disposal by altering organisms to better target and digest waste that would otherwise accumulate. Technology using targeted bioengineered organisms is still very much in its infancy but is considered one of the most promising avenues of environmental biotech. Since the development of this field, researchers have been working on microorganisms that can help in the control of environmental pollutions, while other researchers have been working on engineered organisms to help combat climate change. In 2023, for instance, scientists at the Lawrence National Laboratory in Berkeley announced the discovery of a new pathway utilized by bacteria that might be capable of helping to reduce the carbon emissions from industrial activities. While this technology is only beginning to be explored, this is the kind of genetic engineering that could be utilized alongside a broader shift to renewable energy to combat the threat of climate change.[5]

Another way in which cutting-edge genetic engineering may impact human life is through the creation of new products for consumer use. Examples include bioengineered silk to act as a replacement in the textile industries. While the public debate continues over the benefits/potential risks of bioengineered crops like corn, papayas, and eggplant, newer research is expanding the possibilities for the kinds of organic materials that can be produced through bioengineering and that might benefit consumers in other areas. Genetically modified (GM) cotton and other textiles, for instance, have emerges as one of the new generation of GM products with the potential to alter key industries. However, these technologies have also generated controversy with reports indicating that genetically engineered textiles might have a negative impact on the agriculture industry and could cause further unintended environmental consequences.[6]

Shifting the Future

While genetic technology is still developing, the embrace of genetically engineered products and technologies already indicates the degree to which Americans and citizens around the world have become excited about this pioneering form of genetic engineering. While most of the press coverage has focused on genetic medicine, in the past decade researchers in the biotech industry have expanded their focus and explored new ways that this kind of technological development might impact other industries. This is the kind of research that captivates the public imagination but also reignites fears about the potential problems that may arise. If the use of biotechnology continues to expand, how will societies try to protect workers and entrepreneurs in industries threatened by new technologically produced alternatives to traditional products. Further, how can humanity preserve aspects of human life that are cherished or valued as the biological and sociological realities change under the influence of this emerging technology. These are Some of the broader social/sociological issues that must be addressed as humanity's mastery of genetic engineering continues to develop.[7]

Works Used

Chou, Ann F., et al. "Barriers and Strategies to Integrate Medical Genetics and Primary Care in Underserved Populations: A Scoping Review." *Journal of Community Genetics* 12, no. 3, July 2021.

Dusic, E. J., Tesla Theoryn, Catharine Wang, Elizabeth M. Swisher, and Deborah J. Bowen. "Barriers, Interventions, and Recommendations: Improving the Genetic Testing Landscape." *Frontiers of Digital Health.* Nov. 1, 2022. www.ncbi.nlm.nih.gov/pmc/articles/PMC9665160/. Accessed Aug. 2023.

"Engineered Bacteria Offer a Powerful New Way to Combat Climate Change." *SciTechDaily.* May 10, 2023. scitechdaily.com/engineered-bacteria-offer-a-powerful-new-way-to-combat-climate-change/?expand_article=1.

"Environmental Bioengineering." *MIT School of Bioengineering Sciences and Research*. 2022. mitbio.edu.in/specializations-offered/environmental-engineering/. Accessed Aug. 2023.

Fernández, Clara Rodriguez. "Biotechnology Is Changing How We Make Clothes." *Labiotech*. June 24, 2022. www.labiotech.eu/in-depth/biofabrication-fashion-industry/. Accessed Aug. 2023.

"The Future of Gene Editing." *Columbia University*. Jan. 3, 2020. www.cuimc.columbia.edu/news/future-gene-editing. Accessed Aug. 2023.

Mengstie, Misganaw Asmamaw. "Viral Vectors for the *in Vivo* Delivery of CRISPR Components: Advances and Challenges." *Frontiers Bioengineering and Biotechnology*. 2022.

Notes

1. Chou, et al., "Barriers and Strategies to Integrate Medical Genetics and Primary Care in Underserved Population: A Scoping Review."
2. Dusic, et al., "Barriers, Interventions, and Recommendations: Improving the Genetic Testing Landscape."
3. Mengstie, "Viral Vectors for the *in Vivo* Delivery of CRISRP Components: Advances and Challenges."
4. "Environmental Bioengineering," *MIT School of Bioengineering Sciences & Research*.
5. "Engineered Bacteria Offer a Powerful New Way to Combat Climate Change," *SciTechDaily*.
6. Fernández, "Biotechnology Is Changing How We Make Clothes."
7. "The Future of Gene Editing," *Columbia University*.

The Promise of Gene Editing: So Close and Yet so Perilously Far

By David J. Segal
Frontiers, July 15, 2022

Introduction

On the one hand, it is striking how the promise of genome editing is advancing. Regulatory restrictions have largely eased on genetically engineered crops that carry genome modifications that are similar to spontaneous mutations or those produced by conventional chemical or radiation-based methods (Van Vu et al., 2022). Plants produced by site-directed nuclease type 1 methods (SDN1), for which substitutions and indels are produced only by the action of the nuclease, have been deregulated in many countries. An exception are those countries within the European Union, where, despite being the third largest producer of genetically engineered crops behind China and the USA, SDN1 crops remain subject to the stringent regulations for genetically modified organisms (GMOs). Such stringent regulations are considered to have a dampening effect on agriculture innovation in the EU, and are perhaps similar to the dampening effect of long regulatory delays on the genetic engineering of livestock animals (Van Eenennaam et al., 2021). Since the first report of genetic engineering in livestock animals in 1985, only a single food animal has been commercialized. This is in part due to the USA Food and Drug Administration and their EU counterparts classifying any intentional altered genomic DNA in animals as an investigational new animal drug (INAD) that is not generally recognized as safe. However, there is a growing realization that the current EU policy towards SDN1 crops needs to be updated (Dima et al., 2022), giving hope to the wider use of these directed editing methods that can dramatically accelerate the production of new varieties compared to traditional breeding techniques.

Interestingly, regulations have not hindered innovation in the application of genetic engineering to human health. In fact, this area has been a significant driver of technological advances. Recent publications and scientific meetings, such as the Keystone Symposium on Precision Genome Engineering and the American Society for Gene and Cell Therapy Annual Meeting, highlight the rapid advances in genome editing tools, driven in large part by a sense that new treatments for human disease enabled by these tools are just around the corner. Indeed, by some estimates there are over 100 products using genome editors now in clinical trial (CRISPR

Medicine News), led by companies, such as CRISPR Therapeutics, Intellia Therapeutics, Sangamo Therapeutics, Editas Medicine, Precision Biosciences, Caribou Biosciences, Locus Biosciences, and many others. In the academic sector, Phase 1 of the NIH Somatic Cell Genome Editing Consortium (Saha et al., 2021), which had focused primarily on developing new editors and delivery methods, has led to a Phase 2 that is primarily focused on using these tools to develop treatments up to the stage of submitting an Investigational New Drug (IND) application to the US Food and Drug Administration (FDA).

> **Most gene therapies are not valued based on cost of goods sold; rather, they are value based on calculations such as quality-adjusted-life-years gained by treatment, and how much less expensive the treatment appears compared to lifelong drug or molecular therapy.**

Delivery Challenges

The challenges posed by the demands of eventual human treatments have set several trajectories of innovation. A significant challenge remains the efficient delivery of the editor to the nucleus of the target cell. *Ex-vivo* therapies, such as those for sickle cell disease (SCD) or the production of T-cells carrying a chimeric antigen receptor (CAR T-cells), two of the major indications in this area, can take advantage of a variety of techniques (*e.g.*, electroporation or transfection using lipid nanoparticles [LNP] (Kazemian et al., 2022)) to introduce the editor in various molecular forms (*e.g.*, plasmid DNA, messenger RNA, or ribonucleoprotein complexes [RNP]). The more transient molecular forms, mRNA or RNPs, are generally preferred because plasmid DNA can stimulate innate immune responses in cells, and the long-term expression of the editor can lead to increased off-target effects and immune responses to the editor if expressed in the body. However, the challenges are greater for *in-vivo* therapies. Currently, viral vectors, such as adeno-associated viruses (AAV), remain the most efficient method to deliver editors to cells in the body. However, the limited packaging capacity of AAV poses a formidable constraint. Cas9 from *Staphylococcus aureus* (SaCas9), which is smaller than the more widely used Cas9 from *Streptococcus pyogenes* (SpCas), can just barely fit in an AAV along with an expression cassette for the single-guide RNA (sgRNA). Products such as the EDIT-101 from EDITAS Medicine has used this approach for the treatment of the eye disease Leber Congenital Amaurosis 10, which is now in clinical trial (Maeder et al., 2019). However, larger payloads such tissue-specific promoters, base editors, prime editors or epigenetic editors, are unable to be packaged in a single AAV particle. The challenge of delivering large editor systems has driven several lines of innovation. The Cas9 protein can be split, allowing a two-vector system to deliver N-terminal and C-terminal parts for reassembly in the cell. Though popular in preclinical studies, this approach is generally not favored for clinical applications. A second approach has been the discovery of even smaller Cas9 and Cas12 proteins.

This effort has identified dramatically smaller proteins such as CasMINI (Xu et al., 2021)and CasΦ (Pausch et al., 2020), as well spawned companies dedicated to finding new CRISPR systems in metagenomic data such as Metagenomi and Arbor Biotechnology. A third approach is to abandon AAV in favor of LNP, which have the strong advantages of 1) much larger packaging capacity, 2) far less immune response to the particle (Kenjo et al., 2021), and 3) enable the use of the preferred transient molecular forms of the editor such as mRNA and potentially RNP. NTLA-2001 from Intellia Therapeutics has used this approach for the treatment of the liver disease Transthyretin Amyloidosis, which is now in clinical trials (Gillmore et al., 2021). The major limitation of LNP is that, currently, they are only efficient for de-livery to the liver. However, there are exciting efforts by both academic and industry labs to engineer enhanced transduction capabilities for LNP (Qiu et al., 2021) as well as AAV (Challis et al., 2022), again giving hope to the wider use of these meth-ods to dramatically improve delivery capabilities in the near future.

Editor Challenges

Aside from delivery, another significant challenge is the action of the editor itself. Early concerns about "off-target" editing have largely subsided, perhaps due to a better selection of guide-RNAs, short duration of editor expression, and improved methods for finding off-target events (Giannoukos et al., 2018). However, there are growing concerns about "unexpected events" on-target, such as translocations, very large deletions, loss of entire chromosomal arms, integrations of the viral vec-tor, and chromothripsis (Weisheit et al., 2020; Leibowitz et al., 2021). While these events generally occur in <5% of edited cells, it is essentially a certainty that such events will occur among the million or so cells that are edited in a therapeutic treat-ment. The consequences of these unexpected events remain unclear, and propo-nents would point out that no adverse outcomes such as cancer have been observed in any preclinical or clinical trial. However, the concerns are sufficient to fuel in-creased interest in alternative strategies that do not create double-strand breaks in DNA, such as base, prime, transposon, epigenetic and RNA editors. Beam Thera-peutics, Prime Medicine, Tessera Therapeutics, Integra Therapeutics, SalioGen Therapeutics, and Chroma Medicine are just some of the companies emerging in this space, and are now able to advance towards clinical trials due to the improved delivery systems for large payloads described above. However, many challenges re-main. Base and prime editors allow safe and highly efficient mutation of genes, as in Intellia's knock-out of *TTR* for Transthyretin Amyloidosis mentioned above (Gillmore et al., 2021). However, the promise of base and prime editing to correct mutations in genes is currently less clear, since there are often 10–100 mutated alleles reported for each gene. A change in regulation may again be the answer, if using the same editor with just a different guide-RNA for a particular disorder could be considered safe without requiring full clinical trials to treat each mutant allele. For safer knock-ins, CRISPR-based transposon systems seem poised to overtake nucleases for this function (Klompe et al., 2022; Pallarès-Masmitjà et al., 2021). Nucleases might be considered a "first-generation" technology for targeted insertion of large sequences such as genes because they only create a double-strand break and leave subsequent steps to the many DNA repair pathways in the cell, which is the source of unexpected events. "Second-generation" CRISPR-guided transposon, recombinase and integrase systems perform both break and repair steps, and again benefit from improved delivery systems for their larger payloads. The challenge will be the efficiency and specificity of these systems compared to traditional nuclease-based homology directed repair. For epigenetic editors, the challenge lies in whether life-long changes in target gene expression will require continued expression of the epigenetic editor, or if a short-term treatment can cause a long-term change in the epigenetic information so that expression of the editor is no longer required. The latter might avoid concerns about immune responses to the foreign editor, which will be much more sensitive in humans than in mice, as well as allay concerns about longevity of expression. For RNA editors, the challenge is avoiding collateral damage. It was long known that Cas13 proteins *in vitro* and in bacteria possessed the unusual property of "collateral activity"; once the single-stranded RNA target was cleaved, the enzyme became a non-specific RNase. The collateral activity has been exploited as a system for the sensitive *in-vitro* detection of specific RNAs, such as the SARS-CoV-2 viral RNA by the SHERLOCK system (Gootenberg et al., 2017). Fortuitously, early studies reported the absence of collateral activity in mammalian cells (Abudayyeh et al., 2017; Konermann et al., 2018). Although the mechanism for the loss of this activity was never clear, Cas13 editing was shown to knock down expression of specific mRNAs in mouse models with no apparent adverse effects (Blanchard et al., 2021; Powell et al., 2022). However, more recent studies have found compelling evidence of significant collateral damage of other cellular RNAs in eukaryotic cells (Wang et al., 2019; Ai et al., 2022). This new evidence may lead to a shift away from Cas13 as an RNA nuclease to systems that lack inherent collateral activity, such as the Cas7-11 system (Özcan et al., 2021).

The Most Threatening Challenge

However, there is one challenge that poses an existential threat to the burgeoning field of therapeutic genome editing above all others, and that is cost. As a particularly cautionary example, Glybara, a gene therapy for lipoprotein lipase deficiency, was approved as the first gene therapy in Europe in 2012 but was removed after 2 years on the EU market in 2017 due to poor sales (only a single person was treated outside of a clinical trial). At a cost of ~$1,000,000 per treatment, it was called the most expensive drug in history (Technology Review, 2022), and that was not a price that payers were willing to pay. As we consider the many technical challenges to create genome engineering therapies described above and elsewhere, the realization that the promise of therapy can fail for business reasons is sobering. Yet the first gene therapy approved in the USA, Luxturna, approved in 2017 to treat Leber's congenital amaurosis, costs ~$850,000 to treat both eyes (Darrow, 2019). Zolgensma, approved in the USA in 2019 to treat spinal muscular atrophy, costs ~$2,125,000 per treatment (Garrison et al., 2021). Such prices give pause as to whether gene editing treatments for the 7,000 rare diseases will realistically be accessible to those who urgently need them, or if the limited demand would justify therapeutic development at all. Will the healthcare system overall be able to afford it? Some suggest that, just as the first cars and computers were expensive but are now readily affordable, prices will naturally come down as more gene and gene editing therapies become available. Unfortunately, there are reasons to doubt this scenario. The scenario assumes that prices are primarily based on "cost of goods sold," the direct cost of producing the goods. However, most gene therapies are not valued based on cost of goods sold; rather, they are valued based on calculations such as quality-adjusted-life-years gained by treatment, and how much less expensive the treatment appears compared to life-long drug or molecular therapy (Zimmermann et al., 2019; Dean et al., 2021). For example, the high cost of ~\$15, 000, 000 for treating a hemophilia A patient with factor VIII over the course of their life has been used to justify the comparatively lower cost of Zolgensma (Garrison et al., 2021). Such valuations based on more expensive treatments or monetization of quality-adjusted-life-years are not likely to change due to a reduction in cost of goods. Thus, the promise of gene editing therapies seems so close from a technological perspective, yet so perilously far from a financial perspective.

In conclusion, perhaps for too long the starry-eyed aspirations of technologies developed by academic scientists have been separated from the real-world financial issues involved in translating discoveries into treatments. It is not too soon to ask the question as to whether the current system will serve us well, or if better paradigms could be envisioned. If we are to realize the promise of gene editing therapies, it will likely require academics and industry, foundations and patient advocates, and clinicians and payers to understand each other better. *Frontiers in Genome Editing* is dedicated to giving a forum to all of these stakeholders. We welcome your input and perspective in this important conversation.

References

Abudayyeh, O. O., Gootenberg, J. S., Essletzbichler, P., Han, S., Joung, J., Belanto, J. J., et al. (2017). RNA targeting with CRISPR-Cas13. *Nature* 550, 280–284. doi:10.1038/nature24049.

Ai, Y., Liang, D., and Wilusz, J. E. (2022). CRISPR/Cas13 effectors have differing extents of off-target effects that limit their utility in eukaryotic cells. *Nucleic Acids Res.* 50, e65. doi:10.1093/nar/gkac159.

Blanchard, E. L., Vanover, D., Bawage, S. S., Tiwari, P. M., Rotolo, L., Beyersdorf, J., et al. (2021). Treatment of influenza and SARS-CoV-2 infections via mRNA-encoded Cas13a in rodents. *Nat. Biotechnol.* 39, 717–726. doi:10.1038/s41587-021-00822-w.

Challis, R. C., Ravindra Kumar, S., Chen, X., Goertsen, D., Coughlin, G. M., Hori, A. M., et al. (2022). Adeno-associated virus toolkit to target diverse brain cells. *Annu. Rev. Neurosci.* 45. doi:10.1146/annurev-neuro-111020-100834.

CRISPR Medicine News (2022). Available at: https://crisprmedicinenews.com/clinical-trials/ (accessed 0619, 2022).

Darrow, J. J. (2019). Luxturna: FDA documents reveal the value of a costly gene therapy. *Drug Discov. Today* 24, 949–954. doi:10.1016/j.drudis.2019.01.019.

Dean, R., Jensen, I., Cyr, P., Miller, B., Maru, B., Sproule, D. M., et al. (2021). An updated cost-utility model for onasemnogene abeparvovec (Zolgensma®) in spinal muscular atrophy type 1 patients and comparison with evaluation by the Institute for Clinical and Effectiveness Review (ICER). *J. Mark. Access Health Policy* 9, 1889841. doi:10.1080/20016689.2021.1889841.

Dima, O., Heyvaert, Y., and Inzé, D. (2022). Interactive database of genome editing applications in crops and future policy making in the European Union. *Trends Plant Sci.* S1360-1385.(22), 00140–00146. doi:10.1016/j.tplants.2022.05.002.

Garrison, L. P., Jiao, B., and Dabbous, O. (2021). Gene therapy may not be as expensive as people think: Challenges in assessing the value of single and short-term therapies. *J. Manag. Care Spec. Pharm.* 27, 674–681. doi:10.18553/jmcp.2021.27.5.674.

Giannoukos, G., Ciulla, D. M., Marco, E., Abdulkerim, H. S., Barrera, L. A., Bothmer, A., et al. (2018). UDiTaS™, a genome editing detection method for indels and genome rearrangements. *BMC Genomics* 19, 212. doi:10.1186/s12864-018-4561-9.

Gillmore, J. D., Gane, E., Taubel, J., Kao, J., Fontana, M., Maitland, M. L., et al. (2021). CRISPR-Cas9 *in vivo* gene editing for Transthyretin Amyloidosis. *N. Engl. J. Med.* 385, 493–502. doi:10.1056/NEJMoa2107454.

Gootenberg, J. S., Abudayyeh, O. O., Lee, J. W., Essletzbichler, P., Dy, A. J., Joung, J., et al. (2017). Nucleic acid detection with CRISPR-Cas13a/C2c2. *Science* 356, 438–442. doi:10.1126/science.aam9321.

Kazemian, P., Yu, S.-Y., Thomson, S. B., Birkenshaw, A., Leavitt, B. R., Ross, C. J. D., et al. (2022). Lipid-nanoparticle-based delivery of CRISPR/Cas9 genome-editing components. *Mol. Pharm.* 19, 1669–1686. doi:10.1021/acs.molpharmaceut.1c00916.

Kenjo, E., Hozumi, H., Makita, Y., Iwabuchi, K. A., Fujimoto, N., Matsumoto, S., et al. (2021). Low immunogenicity of LNP allows repeated administrations of CRISPR-Cas9 mRNA into skeletal muscle in mice. *Nat. Commun.* 12, 7101. doi:10.1038/s41467-021-26714-w.

Klompe, S. E., Jaber, N., Beh, L. Y., Mohabir, J. T., Bernheim, A., Sternberg, S. H., et al. (2022). Evolutionary and mechanistic diversity of Type I-F CRISPR-associated transposons. *Mol. Cell* 82, 616–628.e5. doi:10.1016/j.molcel.2021.12.021.

Konermann, S., Lotfy, P., Brideau, N. J., Oki, J., Shokhirev, M. N., Hsu, P. D., et al. (2018). Transcriptome engineering with RNA-targeting type VI-D CRISPR effectors. *Cell* 173, 665–676. e14. doi:10.1016/j.cell.2018.02.033.

Leibowitz, M. L., Papathanasiou, S., Doerfler, P. A., Blaine, L. J., Sun, L., Yao, Y., et al. (2021). Chromothripsis as an on-target consequence of CRISPR-Cas9 genome editing. *Nat. Genet.* 53, 895–905. doi:10.1038/s41588-021-00838-7.

Maeder, M. L., Stefanidakis, M., Wilson, C. J., Baral, R., Barrera, L. A., Bounoutas, G. S., et al. (2019). Development of a gene-editing approach to restore vision loss in Leber congenital amaurosis type 10. *Nat. Med.* 25, 229–233. doi:10.1038/s41591-018-0327-9.

Özcan, A., Krajeski, R., Ioannidi, E., Lee, B., Gardner, A., Makarova, K. S., et al. (2021). Programmable RNA targeting with the single-protein CRISPR effector Cas7-11. *Nature* 597, 720–725. doi:10.1038/s41586-021-03886-5.

Pallarès-Masmitjà, M., Ivančić, D., Mir-Pedrol, J., Jaraba-Wallace, J., Tagliani, T., Oliva, B., et al. (2021). Find and cut-and-transfer (FiCAT) mammalian genome engineering. *Nat. Commun.* 12, 7071. doi:10.1038/s41467-021-27183-x.

Pausch, P., Al-Shayeb, B., Bisom-Rapp, E., Tsuchida, C. A., Li, Z., Cress, B. F., et al. (2020). CRISPR-CasΦ from huge phages is a hypercompact genome editor. *Science* 369, 333–337. doi:10.1126/science.abb1400.

Powell, J. E., Lim, C. K. W., Krishnan, R., McCallister, T. X., Saporito-Magriña, C., Zeballos, M. A., et al. (2022). Targeted gene silencing in the nervous system with CRISPR-Cas13. *Sci. Adv.* 8, eabk2485. doi:10.1126/sciadv.abk2485.

Qiu, M., Li, Y., Bloomer, H., and Xu, Q. (2021). Developing biodegradable lipid nanoparticles for intracellular mRNA delivery and genome editing. *Acc. Chem. Res.* 54, 4001–4011. doi:10.1021/acs.accounts.1c00500.

Saha, K., Sontheimer, E. J., Brooks, P. J., Dwinell, M. R., Gersbach, C. A., Liu, D. R., et al. (2021). The NIH somatic cell genome editing program. *Nature* 592, 195–204. doi:10.1038/s41586-021-03191-1.

Technology Review. (2022). Regalado the world's most expensive medicine is a bust. Available at: https://www.technologyreview.com/s/601165.

Van Eenennaam, A. L., De Figueiredo Silva, F., Trott, J. F., and Zilberman, D. (2021). Genetic engineering of livestock: The opportunity cost of regulatory delay. *Annu. Rev. Anim. Biosci.* 9, 453–478. doi:10.1146/annurev-animal-061220-023052.

Van Vu, T., Das, S., Hensel, G., and Kim, J.-Y. (2022). Genome editing and beyond: What does it mean for the future of plant breeding? *Planta* 255, 130. doi:10.1007/s00425-022-03906-2.

Wang, Q., Liu, X., Zhou, J., Yang, C., Wang, G., Tan, Y., et al. (2019). The CRISPR-cas13a gene-editing system induces collateral cleavage of RNA in glioma cells. *Adv. Sci.* 6, 1901299. doi:10.1002/advs.201901299.

Weisheit, I., Kroeger, J. A., Malik, R., Klimmt, J., Crusius, D., Dannert, A., et al. (2020). Detection of deleterious on-target effects after HDR-mediated CRISPR editing. *Cell Rep.* 31, 107689. doi:10.1016/j.celrep.2020.107689.

Xu, X., Chemparathy, A., Zeng, L., Kempton, H. R., Shang, S., Nakamura, M., et al. (2021). Engineered miniature CRISPR-Cas system for mammalian genome regulation and editing. *Mol. Cell* 81, 4333–4345.e4. e4. doi:10.1016/j.molcel.2021.08.008.

Zimmermann, M., Lubinga, S. J., Banken, R., Rind, D., Cramer, G., Synnott, P. G., et al. (2019). Cost utility of voretigene neparvovec for biallelic RPE65-mediated inherited retinal disease. *Value Health* 22, 161–167. doi:10.1016/j.jval.2018.09.2841.

Print Citations

CMS: Segal, David J. "The Promise of Gene Editing: So Close and Yet so Perilously Far." In *The Reference Shelf: Gene Editing & Genetic Engineering,* edited by Micah L. Issitt, 37–44. Amenia, NY: Grey House Publishing, 2023.

MLA: Segal, David J. "The Promise of Gene Editing: So Close and Yet so Perilously Far." *The Reference Shelf: Gene Editing & Genetic Engineering,* edited by Micah L. Issitt, Grey House Publishing, 2023, pp. 37–44.

APA: Segal, David J. (2023). The promise of gene editing: So close and yet so perilously far. In Micah L. Issitt (Ed.), *The reference shelf: Gene editing & genetic engineering* (pp. 37–44). Amenia, NY: Grey House Publishing. (Original work published 2022)

Engineered Yeast Turn Agricultural Waste into Biofuels

Technology Networks, July 20, 2023

Yeast has been used for thousands of years in the production of beer and wine and for adding fluff and flavor to bread. They are nature's tiny factories that can feed on sugars found in fruit and grains and other nutrients—and from that menu produce alcohol for beverages, and carbon dioxide to make bread rise.

Now researchers at the School of Engineering report making modified yeast that can feed on a wider range of materials, many of which can be derived from agricultural by-products that we don't use—leaves, husks, stems, even wood chips, things often referred to as "waste biomass."

Why Is It Important to Make Yeast That Can Feed on These Agricultural Leftovers?

In recent years, scientists have modified yeast to make other useful products like pharmaceuticals and biofuels. It's a clever way to let nature do our work in a way that does not require toxic chemicals for manufacturing. The technology—referred to as "synthetic biology"—is still young, but looking ahead to a future where biosynthetic production from yeast would operate at a very large scale, we need to feed yeast on something other than what we ourselves need to eat.

Engineering Yeast to Grow on Biomass Sugars

The novel yeast made by the Tufts team can feed on sugars like xylose, arabinose, and cellobiose, which can be extracted from the indigestible woody parts of crops that are often tossed aside after harvesting, such as corn stalks, husks and leaves, and wheat stems. About 1.3 billion tons of this waste biomass is produced each year, providing more than enough sugars to drive a vast industry of yeast biosynthesis.

"If we can get yeast to feed on waste biomass, we can create a biosynthetic industry with a low carbon footprint," said Nikhil Nair, associate professor of chemical and biological engineering. "For example, when we burn biofuels made by yeast, we produce a lot of carbon dioxide, but that carbon dioxide is re-absorbed into crops the following year, which the yeast feed on to make more biofuel, and so on."

Minimal Engineering for Maximum Output

Nair and his team thought that the best chance for efficient consumption of waste biomass sugars might be to modify an existing genetic "dashboard" that the yeast uses to regulate consumption of galactose, a sugar commonly found in dairy products.

> Remodeling yeast to grow on waste biomass sugars sets the stage for improved production of biosynthesized products, which includes drugs such as insulin, human growth hormones, and antibodies.

The dashboard, called a regulon, includes genes for sensors that detect the presence of sugar, and triggers enzymes for the chemical breakdown of sugar so its carbon and oxygen components can be rebuilt into new components. The new components are mostly small molecules and proteins that the yeast itself needs to survive, but they can also be novel products that scientists might have engineered into the yeast.

In an earlier study, the researchers modified the galactose regulon so that the sensor detects the biomass sugar xylose, and triggers enzymes to process xylose instead of galactose.

"Getting yeast to grow on xylose was an important advance," said Sean Sullivan, a graduate student in the Nair lab who co-led the recent study, "but re-engineering different yeast organisms to grow on each biomass sugar is not the best approach. We wanted to design a single yeast organism that can feed off a complete, or nearly complete menu of biomass sugars."

Sullivan made only minimal changes to the regulon already designed for xylose, by changing the sensor protein to more generally accept xylose, arabinose, and cellobiose. Apart from a few more minor changes, the new regulon allowed the yeast organism to grow on these three sugars at rates comparable to yeast grown on native sugars glucose and galactose.

"By using native regulatory networks linked to cell growth and survival, we could take a minimal engineering approach to modifying and optimizing sugar consumption," said Vikas Trivedi, a postdoctoral researcher who co-led the study. "It just so happens that yeast has the machinery to grow on non-native sugars, as long as we adapt sensors and regulons to recognize those sugars."

Improving the Back End of Production

Remodeling yeast to grow on waste biomass sugars sets the stage for improved production of biosynthesized products, which includes drugs such as insulin, human growth hormone, and antibodies. Yeast has also been engineered to produce vaccines by expressing small fragments of virus that stimulate the immune system.

In fact, yeast can be re-engineered to produce natural compounds used to make drugs, which are otherwise difficult to source because they have to be extracted from rare plants. These include scopolamine, used for relieving motion sickness and postoperative nausea, and atropine, used to treat Parkinson's disease patients, and artemensin, used to treat malaria.

Ethanol is a well-known biofuel produced by yeast, but researchers have also engineered the organism to produce other fuels such as isobutanol and isopentanol, which can deliver more energy per liter than ethanol.

Bioengineered yeast can also produce building blocks of bioplastics, such as polylactic acid, which can then be used to make a variety of products, including packaging materials and consumer goods, without having to draw from petroleum sources.

"While the research community continues to innovate yeast to make new products, we are preparing the organism to grow efficiently on agricultural waste biomass, closing a carbon cycle that has so far eluded the manufacturing of fuels, pharmaceuticals and plastics," said Nair.

Print Citations

CMS: Technology Networks. "Engineered Yeast Turn Agricultural Waste into Biofuels." In *The Reference Shelf: Gene Editing & Genetic Engineering,* edited by Micah L. Issitt, 45–47. Amenia, NY: Grey House Publishing, 2023.

MLA: Technology Networks. "Engineered Yeast Turn Agricultural Waste into Biofuels." *The Reference Shelf: Gene Editing & Genetic Engineering,* edited by Micah L. Issitt, Grey House Publishing, 2023, pp. 45–47.

APA: Technology Networks. (2023). Engineered yeast turn agricultural waste into biofuels. In Micah L. Issitt (Ed.), *The reference shelf: Gene editing & genetic engineering* (pp. 45–47). Amenia, NY: Grey House Publishing. (Original work published 2023)

CRISPR/Cas9-Based Gene Drive Could Suppress Agricultural Pests

By Mick Kulikowski
NC State News, June 12, 2023

Targeting the *doublesex* gene resulted in female sterility in numerous experiments as females were unable to lay eggs, says Max Scott, an NC State entomologist who is the corresponding author of a paper in *Proceedings of the National Academy of Sciences* that describes the research.

"This is the first so-called homing gene drive in an agricultural pest that potentially could be used for suppression," Scott said.

Gene drives can preferentially select, change or delete particular traits or characteristics and "drive" those edits through future generations, resulting in a sometimes far greater than 50% chance of passing those changes to progeny.

"Gene drive means biased inheritance," Scott said.

Researchers used a fluorescent red protein to mark the presence of the CRISPR/Cas9 genetic change to the fly's genetic blueprint, or genome. The gene drive systems transmitted that fluorescent protein to 94-99% of progeny, the paper reports.

The researchers also used mathematical modeling to predict how efficiently the gene drive system would suppress a given *D. suzukii* population in laboratory cages. The modeling showed that releasing just one modified fly for every four "wild" flies—those not genetically modified—could tank fly populations within approximately eight to 10 generations.

"Because doublesex is such a conserved gene required for female development in so many fly species, I think the homing gene drive strategy could be used for other pests," Scott said.

Scott and collaborators previously showed success in suppressing *D. suzukii* populations using a strain that produces only males and also used a similar method to reduce lab populations of the New World screwworm fly.

Next steps include contained trial experiments in cages in an NC State greenhouse.

"We're doing small population cage suppression experiments. We're hoping to learn if repeated fly releases with a 1:4 ratio will suppress fly populations in a cage like the modeling suggests," Scott said.

Amarish K. Yadav, an NC State postdoctoral researcher and lead author, Cole Butler, Akihiko Yamamoto, Anandrao A. Patil and Alun L. Lloyd co-authored the paper. The research was supported by Biotechnology Risk Assessment Research program grants 2016-33522-25625, 2020-33522-32317 and 2021-33522-35341 from the U.S. Department of Agriculture's National Institute of Food and Agriculture.

> Gene drives can preferentially select, change or delete particular traits or characteristics and "drive" those edits through future generations. Gene drive means biased inheritance.

Abstract: Genetic-based methods offer environmentally friendly species-specific approaches for control of insect pests. One method, CRISPR homing gene drive that target genes essential for development, could provide very efficient and cost-effective control. While significant progress has been made in developing homing gene drives for mosquito disease vectors, little progress has been made with agricultural insect pests. Here we report the development and evaluation of split homing drives that target the *doublesex* (dsx) gene in *Drosophila suzukii,* an invasive pest of soft-skinned fruits. The drive component, consisting of dsx sgRNA and DsRed genes, was introduced into the female-specific exon of dsx, which is essential for function in females but not males. However, in most strains hemizygous females were sterile and produced the male dsx transcript. With a modified homing drive that included an optimal splice acceptor site, hemizygous females from each of four independent lines were fertile. High transmission rates of the DsRed gene (94-99%) were observed with a line that expressed Cas9 with two nuclear localization sequences from the *D. suzukii* nanos promoter. Mutant alleles of dsx with small in-frame deletions near the Cas9 cut site were not functional and thus would not provide resistance to drive. Finally, mathematical modeling showed that the strains could be used for suppression of lab cage populations of *D. suzukii* with repeated releases at relatively low release ratios (1:4). Our results indicate that the split CRISPR homing gene drive strains could potentially provide an effective means for control of *D. suzukii* populations.

Print Citations

CMS: Kulikowski, Mick. "CRISPR/Cas8-Based Gene Drive Could Suppress Agricultural Pests." In *The Reference Shelf: Gene Editing & Genetic Engineering,* edited by Micah L. Issitt, 48–49. Amenia, NY: Grey House Publishing, 2023.

MLA: Kulikowski, Mick. "CRISPR/Cas8-Based Gene Drive Could Suppress Agricultural Pests." *The Reference Shelf: Gene Editing & Genetic Engineering,* edited by Micah L. Issitt, Grey House Publishing, 2023, pp. 48–49.

APA: Kulikowski, M. (2023). CRISPR/Cas8-based gene drive could suppress agricultural pests. In Micah L. Issitt (Ed.), *The reference shelf: Gene editing & genetic engineering* (pp. 48–49). Amenia, NY: Grey House Publishing. (Original work published 2023)

Baby Without Sperm or Egg? Scientists Develop Synthetic Embryos Using Stem Cells

By Shobhit Gupta
Hindustan Times, June 15, 2023

This major development reportedly dodges the need for eggs or sperms for the creation of human embryos.

In a major breakthrough in the world of In Vitro Fertilisation (IVF) and human reproduction, a team of scientists in the US and the UK has developed the world's first ever synthetic human embryo-like structures using stem cells.

This major development reportedly dodges the need for eggs or sperms for the creation of human embryos.

According to the report by *The Guardian*, these embryos resemble natural embryos in the earliest stages of human development. While they lack a beating heart or the beginnings of a brain, they contain cells that would give rise to the placenta, yolk sac and the embryo. These models would also help scientists understand the impact of genetic disorders and the biological reasons behind recurrent miscarriages.

Professor Magdalena Żernicka-Goetz of the University of Cambridge and the California Institute of Technology, who is leading the research, said, "Our human model is the first three-lineage human embryo model that specifies amnion and germ cells, precursor cells of egg and sperm."

However, the research also raises major ethical and legal consequences as the use of synthetic embryos for clinical purposes is not legally imminent in the UK and most other countries. Implanting them into a patient's womb is currently illegal, and it remains uncertain whether these structures have the potential to progress beyond the earliest stages of development. Scientists at present are allowed to cultivate embryos in the lab for a duration of 14 days.

The research was reportedly inferred on Wednesday in a plenary address at the International Society for Stem Cell Research's annual meeting in US' Boston. However, the deets of the latest findings from the Cambridge-Caltech lab, are yet to be published in a journal paper.

> **These models would also help scientists understand the impact of genetic disorders and the biological reasons behind recurrent miscarriages.**

Goetz and her team, along with a rival team in Israel, had previously described creating model embryo-like structures from mouse stem cells. Those "embryoids" showed the beginnings of a brain, heart and intestinal tract after about eight days of development, the report added.

Goetz told CNN, that the embryo-like structures her lab has created are also the first to have germ cells that would go on to develop into egg and sperm.

Robin Lovell-Badge, head of stem cell biology and developmental genetics at the Francis Crick Institute in London, told the news outlet, "The idea is that if you really model normal human embryonic development using stem cells, you can gain an awful lot of information about how we begin development, what can go wrong, without having to use early embryos for research."

Researchers are in a hope that the synthetic human embryos would shed light on the "black box" of human development.

Print Citations

CMS: Gupta, Shobhit. "Baby without Sperm or Egg? Scientists Develop Synthetic Embryos Using Stem Cells." In *The Reference Shelf: Gene Editing & Genetic Engineering,* edited by Micah L. Issitt, 50–51. Amenia, NY: Grey House Publishing, 2023.

MLA: Gupta, Shobhit. "Baby without Sperm or Egg? Scientists Develop Synthetic Embryos Using Stem Cells." *The Reference Shelf: Gene Editing & Genetic Engineering,* edited by Micah L. Issitt, Grey House Publishing, 2023, pp. 50–51.

APA: Gupta, S. (2023). Baby without sperm or egg? Scientists develop synthetic embryos using stem cells. In Micah L. Issitt (Ed.), *The reference shelf: Gene editing & genetic engineering* (pp. 50–51). Amenia, NY: Grey House Publishing. (Original work published 2023)

The Precious Genes of the World's First Cloned Ferret Could Help Save Her Species

By Joel Goldberg
Science, February 11, 2022

The National Black-footed Ferret Conservation Center in Fort Collins, Colorado, isn't Jurassic Park, but new developments there might sound familiar to fans of the sci-fi classic. This year, the center's sole cloned ferret, a 14-month-old female named Elizabeth Ann, is expected to become the first clone to be bred for the sake of saving her species from extinction.

Three other species have been cloned for conservation: a Przewalski's horse named Kurt, and two types of Southeast Asian cattle under threat, the gaur and the banteng. But Elizabeth Ann is the only clone set to take the next step and breed, an essential step in delivering her unique genes to the shrinking black-footed ferret gene pool.

However, even Elizabeth Ann isn't 100% black-footed ferret. Somatic cell nuclear transfer—the technique used to create her—uses a domestic ferret as a surrogate mother, a process that leaves traces of domestic genes in the cloned offspring. To boost the black-footed ferret genes, scientists hope to one day breed Elizabeth Ann's male offspring with a captive black-footed female, thereby ferreting out any domestic genes. This video (https://www.youtube.com/watch?v=s7M-lrMTFLs) shows how it can be done.

Transcript from YouTube Video

Narrator: Against the backdrop of a Colorado conservation center, the romance of the century is about to unfold. In 2022, researchers will attempt to breed Elizabeth Ann, [Elizabeth Ann barks] the world's first cloned black-footed ferret, with a captive male. If successful, their offspring will infuse the critically endangered species with a splash of genetic diversity. Cloning for the sake of conservation isn't unique to black-footed ferrets.

Another endangered species clone is a Przewalski's horse named Kurt—but Kurt isn't yet ready for breeding. And other earlier clones of banteng and gaur were essentially proofs of concept. They never were bred, limiting their genetic relevance. A lack of genetic diversity is a big problem in re-establishing a species' population. If captive animals are too closely related, they're at risk of inbreeding. That can hobble their reproductive fitness, stymieing family trees. Cloning can help bypass this problem, bringing into play a more genetically distant relative to breed with animals already in captivity. In Elizabeth Ann's case, scientists took cells grown from a skin sample of a long-deceased black-footed ferret—preserved in test tubes at the San Diego Frozen Zoo—to craft an embryo, then transferred it to a

> **Elizabeth Ann is the only clone set to take the next step and breed, an essential step in delivering her unique genes to the shrinking black-footed ferret gene pool.**

domestic ferret that brought the embryo to term. This process—somatic cell nuclear transfer—has been used for decades to duplicate livestock and pets.

Cloning for conservation, meanwhile, has been on the backburner for a few reasons: Breeding practices are less developed for species in need of conservation, and there's less commercial interest in maintaining endangered animals. Also, cloning for conservation has a tricky endgame: to keep a species as genetically pure as possible, biologists prefer the clone's family tree contain only the genes of the endangered species. Animals cloned by somatic cell nuclear transfer still carry their domestic mothers' mitochondrial DNA. This means the offspring of a cloned animal—such as Elizabeth Ann—stand to bring domestic genes into the endangered population. But because mitochondrial DNA is inherited from an animal's mother, these domestic genes can be ferreted out by breeding the male offspring of clones with captive females. If successful, the process could make an endangered population more resilient, and eventually lead to self-sustained growth. Just over a month removed from her first birthday, Elizabeth Ann remains a marvel in her own right. If she finds her captive male suitor to be just as marvelous, their once-in-a-lifetime romance could propel their species for lifetimes to come.

Print Citations

CMS: Goldberg, Joel. "The Precious Genes of the World's First Cloned Ferret Could Help Save Her Species." In *The Reference Shelf: Gene Editing & Genetic Engineering,* edited by Micah L. Issitt, 52–53. Amenia, NY: Grey House Publishing, 2023.

MLA: Goldberg, Joel. "The Precious Genes of the World's First Cloned Ferret Could Help Save Her Species." *The Reference Shelf: Gene Editing & Genetic Engineering,* edited by Micah L. Issitt, Grey House Publishing, 2023, pp. 52–53.

APA: Goldberg, J. (2023). The previous genes of the world's first cloned ferret could help save her species. In Micah L. Issitt (Ed.), *The reference shelf: Gene editing & genetic engineering* (pp. 52–53). Amenia, NY: Grey House Publishing. (Original work published 2022)

3
Genetic Medicine

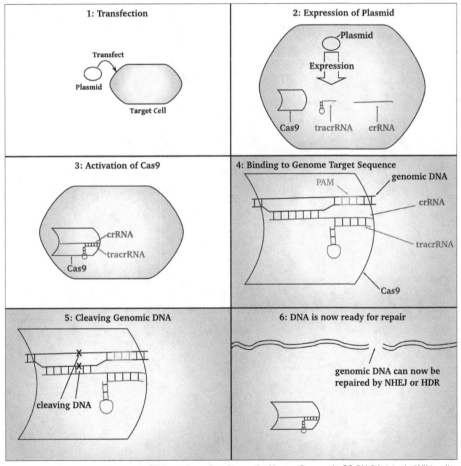

Image by Nielsrca based on figures in *Nature Protocols*, CC BY-SA 4.0, via Wikipedia.

The CRISPR method is one technique used in genetic medicine. Above, a diagram of DNA cleaving by CRISPR-Cas9.

Gene Therapy and Medical Genetics

Genes are the primary unit of heredity, allowing reproducing humans and other organisms to pass on their physical traits to their offspring, but this is only one of the many roles that genes play in biology.

It is estimated that humans have between 20,000 and 25,000 genes, making up the human genome, and copies of these genes are present within human cells. During reproduction, the parent or parents pass on genes that contain instructions for how to create a body. The recombination of genes from parents is what creates unique characteristics in offspring and also explains why children typically share some traits with their parents.

In addition to this reproductive function, as a vector for heritable characteristics, genes also contain instructions for creating proteins and other substances needed to keep the body working. Proteins are complex molecules that play many different roles in the body and do most of the "work" that goes on within cells. When a human breathes, for instance, it is proteins that allow for the exchange of gases and atoms that deliver oxygen to the tissues of the body. When a person "thinks," proteins in the brain actively facilitate the firing of neurons. There is a special class of complex molecules known as "enzymes" that can be either proteins or ribonucleic acid (RNA) molecules and have proven to be among the most essential substances in the body. Enzymes facilitate chemical reactions and so they are needed to regulate and stimulate body chemistry. It is a person's genes that contain the instructions for making proteins and RNA and so it is the genes that determine not only how a person's body develops, but also how the body functions.

Because genes are involved in both heredity and the day-to-day function of biological bodies, genes play a major role in overall health. Genetic abnormalities can cause disease or physical problems that can greatly impact a person's life. Many of the best-known human diseases, like diabetes, cancer, and cardiovascular disease, have been linked to genetic factors and scientists believe it is likely that virtually all diseases and disorders have some level of genetic influence. Over the past century, scientists have begun to better understand how genes work and how they contribute to health and this led to the birth of a new field of medical science, known as "genetic medicine."[1]

Heritable Genetic Disorders

Medical specialists in the ancient world already understood that some kinds of health issues can be passed down from parents to offspring. It was observed that some blind people, for instance, could pass on blindness to their children, though not all vision loss is attributable to genetic factors.[2] Observing how traits like baldness, hearing loss, vision loss, and other obvious physical differences tend to occur within families was the first indication of heritable health issues, but it wasn't until humanity learned about genes and how they work that medical professionals began to understand the complicated links between heritability and health.

Over the past fifty years, the science of heritable disease has grown and developed rapidly and humanity has discovered more about how susceptibility to disease is related to genetics. One of the innovations that has emerged from this research is the potential to test individuals for genetic factors linked to disease. Genetic testing is available for certain developmental diseases and for certain cancers, though these kinds of genetic tests are expensive and may not be available to all providers and patients.[3] The potential to test for susceptibility has raised complicated questions about how best to utilize this information. If, to give an example, an individual submits to genetic testing and learns that they might pass on genes related to a certain disease, should the person avoid reproducing to prevent them from potentially passing on a disease or disorder? This is one of the ethical issues that frequently appears in debates about genetic medicine.

In most cases, diseases and disorders are not entirely "genetic" in nature but are influenced by both genetic and environmental factors. There are genetic characteristics that might make it more likely for a person to develop a certain disease or disorder and this is how certain diseases and disorders can be heritable. Because there are heritable genetic components to certain diseases, this raises the question of whether it is possible to prevent diseases by treating children for genetic disorders prior to birth or by eliminating genetic abnormalities that might lead to the development of a gene-linked medical issue. Over the past twenty years, scientists have been working towards "prenatal gene therapy" in which medical professionals could use gene therapy to treat a developing fetus prior to birth, which might mean preventing a disease before it develops. While this kind of genetic medicine is still very much in its infancy, some medical scientists believe that this may be one of the most important techniques of the future.[4]

Gene Therapy Beyond Heredity

While one of the foci of genetic testing and genetic medicine is to address heritable disease, gene therapy can also be used to treat individuals of any age struggling with disorders or diseases related to the function of genes within the body. As discussed, genes code for enzymes and other kinds of proteins and these molecules are essential to the proper function of the body. A mutation in a gene might mean that a person fails to produce an important protein or other molecule. A number of devastating diseases have been linked to malfunctions in the genes and one aspect of genetic medicine involves identifying and potentially treating diseases and disorders related to malfunctions in gene expression.

There are two major methods of gene therapy utilized in genetic medicine. The first technique is known as "ex vivo," which means that the treatment takes place, initially, outside of the body. In ex vivo treatment, cells are removed from a patient's body, and altered genetic material is introduced to the cells using a "vector" which is a virus or other biological mechanism that can be used to introduce genes into a system. These cells are then returned to the patient's body. This method of gene therapy has been used as a treatment for sickle cell disease, a blood disorder in which an individual develops mutated red blood cells that cannot effectively function and is related to a malfunctioning gene. The other main type of gene therapy is "In vivo" therapy in which genetically modified (GM) material is injected into the bloodstream with a vector, after which the gene incorporates into cells and alters the way that genes are expressed. In vivo gene therapy has been used to treat disorders like hemophilia, an inherited blood disorder that interferes with an individual's ability to form blood clots.[5]

An Emerging Field

Genetic medicine is very much an emerging field and it is only in the past twenty years that gene therapies have reached a functional stage. In the 2020s, scientists have been able to utilize genetic manipulation tools like Clustered Regularly Interspaced Short Palindromic Repeats (CRISPR), as well as viral vectors, virus components used to deliver and incorporate genetic material into a person's genes, to create functional medical interventions for some of the most ancient and problematic human conditions. However, in many cases genetic medicine is still experimental and few patients are able to access genetic treatments.

Among the most recent developments in the field of genetic medicine is the use of genetic editing to avoid immune system rejection in transplant medicine. In addition, researchers have had breakthroughs utilizing modified bacteria and viruses to deliver medical treatments and to better detect genetic disorders. Experts in the field have also noted that it is likely that the incorporation of artificial intelligence (AI) into genetic research could dramatically alter the practice of gene therapy in the near future.

While many Americans and citizens around the world welcome genetic medicine as a way to combat persistent, heritable or genetic diseases, genetic treatments also remain controversial. Some feel that altering a person's genes is an "unnatural" form of medicine, while others worry that too little is known about genetics and genetic disorders to safely use genetic treatments. One of the biggest challenges for supporters of gene therapy and other forms of genetic medicine is to combat barriers preventing members of the public from accessing and utilizing genetic treatments and also to conduct long-term testing to evaluate the safety of gene therapy techniques, which will also increase public confidence in the use of these new medical tools.

Works Used

Fliesler, Nancy. "After Decades of Evolution, Gene Therapy Arrives." *Boston Children's Hospital*. Dec. 22, 2020. answers.childrenshospital.org/gene-therapy-history/. Accessed Aug. 2023.

"Genetic Testing." *CDC*. Genomics & Precision Health. www.cdc.gov/genomics/gtesting/genetic_testing.htm. Accessed Aug. 2023.

Muacevic, Alexander, and John R. Adler. "Intrauterine Fetal Gene Therapy: Is That the Future and Is That Future Now?" *Cureus*. Feb. 23, 2022.

Raju, Leela. "Is Blindness Genetic? What to Know." *Medical News Today*. Apr. 11, 2023. www.medicalnewstoday.com/articles/is-blindness-genetic. Accessed Aug. 2023.

"What Is a Gene?" *Medline Plus*. medlineplus.gov/genetics/understanding/basics/gene/. Accessed Aug. 2023.

Notes

1.   "What Is a Gene?" *Medline Plus*.
2.   Raju, "Is Blindness Genetic? What to Know."
3.   "Genetic Testing," *CDC*.
4.   Muacevic and Adler, "Intrauterine Fetal Gene Therapy: Is That the Future and Is That Future Now?"
5.   Fliesler, "After Decades of Evolution, Gene Therapy Arrives."

Edits to a Cholesterol Gene Could Stop the Biggest Killer on Earth

By Antonio Regalado
MIT Technology Review, July 12, 2022

A volunteer in New Zealand has become the first person to undergo DNA editing in order to lower their blood cholesterol, a step that may foreshadow wide use of the technology to prevent heart attacks.

The experiment, part of a clinical trial by the US biotechnology company Verve Therapeutics, involved injecting a version of the gene-editing tool CRISPR in order to modify a single letter of DNA in the patient's liver cells.

According to the company, that tiny edit should be enough to permanently lower a person's levels of "bad" LDL cholesterol, the fatty molecule that causes arteries to clog and harden with time.

The patient in New Zealand had an inherited risk for extra-high cholesterol and was already suffering from heart disease. However, the company believes the same technique could eventually be used on millions of people in order to prevent cardiovascular disease.

"If this works and is safe, this is the answer to heart attack—this is the cure," says Sekar Kathiresan, a gene researcher who started Verve three years ago and is the company's CEO.

It's been 10 years since scientists developed CRISPR, a technology for making targeted changes to the DNA in cells, but until now the method has been tried only on people suffering from rare diseases like sickle-cell anemia, and only as part of exploratory trials.

If Verve's experiment works, it could signal far wider use of gene editing to prevent common conditions. Large swaths of the world's population have LDL that is too high, but many people can't get it under control. Worldwide, more people die of atherosclerotic cardiovascular disease than from anything else.

"Of all the different genome editing ongoing on the clinic, this one could have the most profound impact because of the number of people who could benefit," says Eric Topol, a cardiologist and researcher at Scripps Research.

Some doctors believe lowering LDL aggressively, and keeping it low throughout life, could essentially prevent people from dying of cardiovascular disease. That is the view of Eugene Braunwald, a physician at Brigham & Women's Hospital in Boston who is also an advisor to Verve.

"The lower the LDL, the better," says Braunwald. "You can't have too low an LDL. The problem is how do you get it down?"

A strict diet, like one where you avoid eggs, meat, and even olive oil, can help. But few people succeed in sticking to it. Then come statins, the most widely prescribed medicine in the US. These pills can cut a person's LDL in half, but some can't handle the side effects, and some find even taking a once-a-day pill hard to manage.

Some newer biotech drugs involve injections twice a month, or even just twice a year. These drugs are quite powerful, and Braunwald recently speculated what would occur if they were given widely as a public health intervention, not unlike an annual flu vaccine. "I calculate it if you give it starting at age 30, you will live to 100 without coronary artery disease," he says.

However, those drugs aren't yet widely used. They remain costly, are still inconvenient, and insurers balk at paying. "So gene editing is the big stick, because it's one and done. You don't ever have to come back," says Braunwald. "It's a very big deal, because atherosclerotic cardiovascular disease is the most common cause of death in the industrialized world, and LDL is the primary reason."

In New Zealand, where Verve's clinical trial is taking place, doctors will give the gene treatment to 40 people who have an inherited form of high cholesterol known as familial hypercholesterolemia, or FH. People with FH can have cholesterol readings twice the average, even as children. Many learn they have a problem only when they get hit with a heart attack, often at a young age.

The study also marks an early use of base editing, a novel adaptation of CRISPR that was first developed in 2016. Unlike traditional CRISPR, which cuts a gene, base editing substitutes a single letter of DNA for another.

The gene Verve is editing is called *PCSK9*. It has a big role in maintaining LDL levels and the company says its treatment will turn the gene off by introducing a one-letter misspelling.

Before starting Verve, Kathiresan was a geneticist working at the Broad Institute in Cambridge, Massachusetts, looking for inherited causes of heart disease. He started Verve after his brother, Senthil, was struck down suddenly by a heart attack; base editing, he thought, could be a way to prevent such tragedies.

One reason Verve's base-editing technique is moving fast is that the technology is substantially similar to mRNA vaccines for covid-19. Just like the vaccines, the treatment consists of genetic instructions wrapped in a nanoparticle, which ferries everything into a cell.

While the vaccine instructs cells to make a component of the SARS-CoV-2 virus, the particles in Verve's treatment carry RNA directions for a cell to assemble and aim a base-editing protein, which then modifies that cell's copy of *PCSK9*, introducing the tiny mistake.

In experiments on monkeys, Verve found that the treatment lowered bad cholesterol by 60%. The effect has lasted more than a year in the animals and could well be permanent.

The human experiment could entail some risk. Nanoparticles are somewhat toxic, and there have been reports of side effects, like muscle pain, in people taking other drugs to lower *PCSK9*. And whereas treatment with ordinary drugs can be discontinued if problems come up, there's as yet no plan to undo gene editing once it's performed.

The main cause of death in the world is also the first common problem that gene editing can address.

So far, the few gene therapies on the market all cost hundreds of thousands of dollars—even as much as $2 million. But Verve's should be much cheaper, especially if used widely. One reason is that whereas other gene therapies use specially prepared viruses to carry genes, nanoparticles are made in a chemical process that's more practical to scale up.

"The pandemic and the emergent need for vaccines [created] large-scale manufacturing capacity," says Kiran Musunuru, a gene-editing expert at the University of Pennsylvania who cofounded Verve. That capacity "can be easily repurposed for genetic therapy," he says, and "of course, abundant capacity means reduced prices."

Musunuru says people are even thinking about "booster shots" in case the first round of gene editing isn't complete, or in order to knock out other cholesterol genes and deepen the effect on LDL.

It's a stroke of luck for Verve's founders that the main cause of death in the world is also the first common problem that gene editing can address. Kathiresan, who takes a statin to keep his LDL low, says he thinks gene editing for cholesterol has the potential to be a life-extension treatment.

"The number one cause of mortality in the world is heart attack," he says. "If you are going to give a medicine that makes you avoid a heart attack, people are going to live longer."

Print Citations

CMS: Regalado, Antonio. "Edits to a Cholesterol Gene Could Stop the Biggest Killer on Earth." In *The Reference Shelf: Gene Editing & Genetic Engineering,* edited by Micah L. Issitt, 61–63. Amenia, NY: Grey House Publishing, 2023.

MLA: Regalado, Antonio. "Edits to a Cholesterol Gene Could Stop the Biggest Killer on Earth." *The Reference Shelf: Gene Editing & Genetic Engineering,* edited by Micah L. Issitt, Grey House Publishing, 2023, pp. 61–63.

APA: Regalado, A. (2023). Edits to a cholesterol gene could stop the biggest killer on earth. In Micah L. Issitt (Ed.), *The reference shelf: Gene editing & genetic engineering* (pp. 61–63). Amenia, NY: Grey House Publishing. (Original work published 2022)

He Inherited a Devastating Disease: A CRISPR Gene-Editing Breakthrough Stopped It

By Rob Stein
NPR, June 26, 2021

Patrick Doherty had always been very active. He trekked the Himalayas and hiked trails in Spain.

But about a year and a half ago, he noticed pins and needles in his fingers and toes. His feet got cold. And then he started getting out of breath any time he walked his dog up the hills of County Donegal in Ireland where he lives.

"I noticed on some of the larger hill climbs I was getting a bit breathless," says Doherty, 65. "So I realized something was wrong."

Doherty found out he had a rare, but devastating inherited disease—known as transthyretin amyloidosis—that had killed his father. A misshapen protein was building up in his body, destroying important tissues, such as nerves in his hands and feet and his heart.

Doherty had watched others get crippled and die difficult deaths from amyloidosis.

"It's terrible prognosis," Doherty says. "This is a condition that deteriorates very rapidly. It's just dreadful."

So Doherty was thrilled when he found out that doctors were testing a new way to try to treat amyloidosis. The approach used a revolutionary gene-editing technique called CRISPR, which allows scientists to make very precise changes in DNA.

"I thought: Fantastic. I jumped at the opportunity," Doherty says.

On Saturday, researchers reported the first data indicating that the experimental treatment worked, causing levels of the destructive protein to plummet in Doherty's body and the bodies of five other patients treated with the approach.

"I feel fantastic," Doherty says. "It's just phenomenal."

The advance is being hailed not just for amyloidosis patients but also as a proof-of-concept that CRISPR could be used to treat many other, much more common diseases. It's a new way of using the innovative technology.

"This is a major milestone for patients," says Jennifer Doudna of the University of California, Berkeley, who shared a Nobel Prize for her work helping develop CRISPR.

"While these are early data, they show us that we can overcome one of the biggest challenges with applying CRISPR clinically so far, which is being able to deliver it systemically and get it to the right place," Doudna says.

> **CRISPR has been shown to help patients suffering from devastating blood disorders sickle cell disease and beta thalassemia; doctors are trying to use it to treat cancer and to restore vision to people blinded by a rare genetic disorder.**

CRISPR has already been shown to help patients suffering from the devastating blood disorders sickle cell disease and beta thalassemia. And doctors are trying to use it to treat cancer and to restore vision to people blinded by a rare genetic disorder.

But those experiments involve taking cells out of the body, editing them in the lab, and infusing them back in or injecting CRISPR directly into cells that need fixing.

The study Doherty volunteered for is the first in which doctors are simply infusing the gene-editor directly into patients and letting it find its own way to the right gene in the right cells. In this case, it's cells in the liver making the destructive protein.

"This is the first example in which CRISPR-Cas9 is injected directly into the bloodstream—in other words systemic administration—where we use it as a way to reach a tissue that's far away from the site of injection and very specifically use it to edit disease-causing genes," says John Leonard, the CEO of Intellia Therapeutics, which is sponsoring the study.

Doctors infused billions of microscopic structures known as nanoparticles carrying genetic instructions for the CRISPR gene-editor into four patients in London and two in New Zealand. The nanoparticles were absorbed by their livers, where they unleashed armies of CRISPR gene-editors. The CRISPR editor homed in on the target gene in the liver and sliced it, disabling production of the destructive protein.

Within weeks, the levels of protein causing the disease plummeted, especially in the volunteers who received a higher dose. Researchers reported at the Peripheral Nerve Society Annual Meeting and in a paper published in *The New England Journal of Medicine*.

"It really is exciting," says Dr. Julian Gillmore, who is leading the study at the University College London, Royal Free Hospital.

"This has the potential to completely revolutionize the outcome for these patients who have lived with this disease in their family for many generations. It's decimated some families that I've been looking after. So this is amazing," Gillmore says.

The patients will have to be followed longer, and more patients will have to be treated, to make sure the treatment's safe, and determine how much it's helping, Gillmore stresses. But the approach could help those struck by amyloidosis that isn't inherited, which is a far more common version of the disease, he says.

Moreover, the promising results potentially open the door for using the same approach to treatment of many other, more common diseases for which taking cells out of the body or directly injecting CRISPR isn't realistic, including heart disease, muscular dystrophy and brain diseases such as Alzheimer's.

"This is really opening a new era as we think about gene-editing where we can begin to think about accessing all kinds of different tissue in the body via systemic administration," Leonard says.

Other scientists who are not involved in the research agree.

"This is a wonderful day for the future of gene-editing as a medicine," agree Fyodor Urnov, a professor of genetics at the University of California, Berkeley. "We as a species are watching this remarkable new show called: our gene-edited future."

Doherty says he started feeling better within weeks of the treatment and has continued to improve in the weeks since then.

"I definitely feel better," he told NPR. "I'm speaking to you from upstairs in our house. I climbed stairs to get up here. I would have been feeling breathless. I'm thrilled."

Print Citations

CMS: Stein, Rob. "He Inherited a Devastating Disease: A CRISPR Gene-Editing Breakthrough Stopped It." In *The Reference Shelf: Gene Editing & Genetic Engineering*, edited by Micah L. Issitt, 64–66. Amenia, NY: Grey House Publishing, 2023.

MLA: Stein, Rob. "He Inherited a Devastating Disease: A CRISPR Gene-Editing Breakthrough Stopped It." *The Reference Shelf: Gene Editing & Genetic Engineering*, edited by Micah L. Issitt, Grey House Publishing, 2023, pp. 64–66.

APA: Stein, R. (2023). He inherited a devastating disease: A CRISPR gene-editing breakthrough stopped it. In Micah L. Issitt (Ed.), *The reference shelf: Gene editing & genetic engineering* (pp. 64–66). Amenia, NY: Grey House Publishing. (Original work published 2021)

Organ Transplants from Pigs: Medical Miracle or Pandemic in the Making?

By J. Roger Jacobs
The Conversation, February 27, 2022

Three out of four new diseases are zoonotic, meaning they have evolved to infect new host species. For example, a mutated bird-flu virus may jump from wild birds to free-range domestic poultry and then to humans who are in contact with poultry. Similar pathways have led to infection by the pathogens that cause Ebola, Zika, HIV, Lyme disease and likely COVID-19.

If a new medical technology increased the risk of a new zoonotic pandemic—however marginally—how would society decide the balance of risk and benefit? If you needed new lungs that were only available in another country, would a health prohibition on the transplant in your own country stop you?

New developments in organ transplant technology may have streamlined a pathway for new zoonotic diseases, but the biotechnology innovators and medical research institutes have not engaged the public on the risks. Failing to do so may jeopardize the potential of a promising therapy.

Xenotransplantation

Over 4,400 Canadians are waitlisted for the lifesaving transplant of a new kidney, liver or lung. In 2019, 250 died waiting. In the United States and elsewhere, the supply gap is more extreme and high hopes ride on xenotransplantation: the transplanting of cells, tissues or organs from animals.

Pre-clinical trials of organ transplants from pigs have addressed the technical barriers to xenotransplantation, reducing the likelihood of rejection. Last summer, Maryland School of Medicine surgeons reported the 31-day survival of a baboon after receiving a lung from a genetically modified pig.

Weeks later, a team at New York University transplanted a kidney from a genetically modified pig into a brain-dead person. In December 2021, surgeons at Maryland School of Medicine transplanted a genetically modified pig heart into a living 57-year-old man.

All projects were approved under U.S. Food and Drug Administration (FDA) regulations, and corporate funding was supplemented by the U.S. National Institutes of Health. The next step with the FDA is to approve clinical trials. Normalization of xenotransplantation could happen before there is informed public acceptance of the benefits and risks.

A Potential Zoonotic Pathway

As a developmental geneticist, it has been exciting to track these advances. The revolution in designer gene editing (known as CRISPR-Cas9) makes this stunning progress possible. CRISPR allows molecules on the surface of pig cells to be modified so the human immune system will not trigger tissue rejection.

Immense pressure to resolve the growing organ shortage for transplantation may tempt the biotechnology business and public regulators to be insufficiently critical.

To prevent human transplant recipients from being infected with pig retroviruses (viruses that can integrate their genetic material into the host's cells), the retroviruses hiding in the pig genome have been removed by CRISPR editing. The risk of transferring a disease directly from a genetically modified donor pig to the human host is negligible.

However, disease-free transplanted pig organs could become infected after transplantation. Zoonotic bacteria and viruses enter hosts most readily through the delicate surfaces of the respiratory tract, as with COVID-19. Living pig cells in a transplanted lung could readily be infected by an inhaled pig virus, including a novel virus from a wild animal host that has evolved to infect pigs.

After entering the human body, a replicating zoonotic virus could generate millions of mutations a day, because their mechanism for gene copying is naturally error prone. A pig virus replicating in a lung transplanted into a human could produce variants that may be capable of recognizing and infecting human cells. Although likely a rare event, it is not impossible that this could trigger a new zoonotic pandemic.

Risk, Fear and Polarization

The scenario described above could evoke risk and fear from a complex new medical technology. It parallels the thinking involved in vaccine hesitancy or the distrust of genetically modified foods. Both are well anchored in today's political culture. In both cases, citizens increasingly demand prior consent and the choice to opt out—despite possible risks to public health. Vaccine hesitancy has increased the death toll from COVID-19 and delayed economic recovery from the pandemic.

In contrast, distrust of the industrialization of food has discouraged introduction of genetically modified foods that enhance nutrition or sustain agricultural productivity in a warming climate. Consumers question whether genetically modified organisms (GMOs) exist for public benefit or for corporate profit.

Increasingly, health issues such as vaccination, vaping or genetic testing generate highly polarized platforms for misinformation, debate and political leverage. Social media algorithms amplify extreme positions and elicit strong emotional reactions at the expense of the middle ground. When communications from the scientific community are reactive, poorly targeted or unintelligible to the average person, the influence of science in the policy process is diminished.

In 2022, progress in xenotransplant technology makes good news stories. Immense pressure to resolve the growing organ shortage for transplantation may tempt the biotechnology business and public regulators to be insufficiently critical as they seek permission to proceed with clinical studies. They must prepare for the nature and scale of backlash from those tired of experts and mistrustful of corporate motivation and institutional authority.

Concern about zoonosis from transplants was voiced over twenty years ago, long before CRISPR transformed the field. Since then, there appear to be no hard facts or even a call for research on zoonotic infection through xenotransplants after transplantation. Bioethicists are flagging the issue now, but the silence about xenotransplant zoonosis from biotechnology corporations and their affiliated preclinical research institutes leaves an open door to a narrative motivated by skepticism and distrust. It is incumbent on them to lead a public dialogue on managing the risk of novel zoonotic diseases arising from infection after transplantation.

Print Citations

CMS: Jacobs, J. Roger. "Organ Transplants from Pigs: Medical Miracle or Pandemic in the Making?." In *The Reference Shelf: Gene Editing & Genetic Engineering,* edited by Micah L. Issitt, 67–69. Amenia, NY: Grey House Publishing, 2023.

MLA: Jacobs, J. Roger. "Organ Transplants from Pigs: Medical Miracle or Pandemic in the Making?." *The Reference Shelf: Gene Editing & Genetic Engineering,* edited by Micah L. Issitt, Grey House Publishing, 2023, pp. 67–69.

APA: Jacobs, J. R. (2023). Organ transplants from pigs: Medical miracle or pandemic in the making?. In Micah L. Issitt (Ed.), *The reference shelf: Gene editing & genetic engineering* (pp. 67–69). Amenia, NY: Grey House Publishing. (Original work published 2022)

Using Gene Editing to Fight Deadly Genetic Diseases

By Karen Feldscher
Harvard T.H. Chan, November 30, 2022

November 30, 2022—Cutting-edge gene editing techniques hold enormous promise for tackling devastating diseases such as sickle cell disease, Huntington's disease, and heart disease, according to experts.

At the 16th annual Program in Quantitative Genomics (PQG) conference, a two-day event held in early November and hosted by Harvard T.H. Chan School of Public Health, a dozen speakers spoke about recent and upcoming research on therapeutics and technologies targeting specific genetic mutations that cause disease. About 180 participants from around the world attended the virtual conference.

"The mutations in our genomes cause about 7,000 known genetic diseases that collectively affect hundreds of millions of people and their families," said David Liu, Richard Merkin Professor and Director of the Merkin Institute of Transformative Technologies in Healthcare, Broad Institute of MIT and Harvard and Thomas Dudley Cabot Professor of the Natural Sciences at Harvard, one of three keynote speakers at the conference. "So a longstanding goal of the life sciences has been to develop the ability to … correct as many of these mutations as possible, so that we can study or treat the broadest possible range of the resulting diseases."

Other keynote speakers included Sekar Kathiresan, chief executive officer and founder of Verve Therapeutics; and Beverly Davidson, professor of pathology and laboratory medicine, Perelman School of Medicine, University of Pennsylvania/Children's Hospital of Philadelphia.

Kathiresan spoke about interventions that can help patients with a high genetic risk of heart attack. He noted that people with low levels of LDL cholesterol—so-called "bad cholesterol"—rarely get heart attacks. He described Verve Therapeutics' development of a one-time intravenous drug that, in animal models, has been successful in turning off a gene in the liver called PCSK9, which ultimately leads to lower LDL. The treatment is currently undergoing human trials.

"Ultimately, this medicine could truly be a preventive medicine, basically treating patients … to make sure they don't get their first heart attack," Kathiresan said.

Davidson discussed how gene editing can fight Huntington's disease. She described research into the use of a non-pathogenic virus called AAV (adeno-associated virus) to deliver gene therapy to quell the effects of the mutation that causes Huntington's.

Liu described two types of gene editing that can make precise changes in genes: base editing, which can change a single DNA letter; and prime editing, which can safely delete or repair long lengths of disease-causing DNA or insert DNA to repair dangerous mutations. "The vast majority of known genetic diseases require precise target gene correction as opposed to gene disruption or deletion," he explained. He and colleagues used base editing, for example, to alter the mutant gene that causes progeria—the rapid aging disease—significantly reducing the disease's effects in animal models.

> **New gene editing technologies give us hope that one day we may no longer be so beholden to the misspellings in our DNA.**

New gene editing technologies, he noted "give us hope that one day we may no longer be so beholden to the misspellings in our DNA."

Other speakers at the conference covered topics such as gene editing for blood disorders and using genomics to discover new cancer drugs. The conference also featured a virtual poster session space that enabled participants to create avatars that could "walk" from poster to poster in a virtual room and interact with poster presenters and others.

Luca Pinello, associate professor at Massachusetts General Hospital and Harvard Medical School, chaired the seven-member organizing committee for the conference and introduced the event; Xihong Lin, professor of biostatistics at Harvard Chan School, also served on the organizing committee.

Print Citations

CMS: Feldscher, Karen. "Using Gene Editing to Fight Deadly Genetic Diseases." In *The Reference Shelf: Gene Editing & Genetic Engineering*, edited by Micah L. Issitt, 70–71. Amenia, NY: Grey House Publishing, 2023.

MLA: Feldscher, Karen. "Using Gene Editing to Fight Deadly Genetic Diseases." *The Reference Shelf: Gene Editing & Genetic Engineering*, edited by Micah L. Issitt, Grey House Publishing, 2023, pp. 70–71.

APA: Feldscher, K. (2023). Using gene editing to fight deadly genetic diseases. In Micah L. Issitt (Ed.), *The reference shelf: Gene editing & genetic engineering* (pp. 70–71). Amenia, NY: Grey House Publishing. (Original work published 2022)

COVID Testing Led to New Techniques of Disease Diagnosis: Progress Mustn't Stop Now

Angelika Loots
The Conversation, March 24, 2023

In March 2020, weeks before the World Health Organization (WHO) declared CO-VID-19 a pandemic, its director-general Tedros Adhanom Ghebreyesus delivered a speech in which he emphasised the importance of testing:

> … the most effective way to prevent infections and save lives is breaking the chains of transmission. And to do that, you must test and isolate. You cannot fight a fire blindfolded. And we cannot stop this pandemic if we don't know who is infected. We have a simple message for all countries: test, test, test.

The pandemic exposed critical shortcomings of existing diagnostic techniques. It revealed an urgent need for tests that are faster, simpler, cheaper and more scalable than existing methods, and just as accurate.

Three years on, the global face of diagnostics has changed. New techniques of disease diagnosis have been developed that can be applied to other emerging zoonotic pathogens such as "disease X"—a hypothetical infectious disease that has the potential to develop into a pandemic.

As a molecular scientist with a keen interest in veterinary disease diagnostics, I have closely followed developments in the diagnostic space since the start of the pandemic. These emerging technologies, together with conventional tests, have the potential to overcome bottlenecks in the current diagnostic procedures. By incorporating these tests into a country's healthcare system, clinicians and policy makers are better equipped to practise precision medicine and to react to potential outbreaks.

How the Tests Changed

The first diagnostic tests for SARS-CoV-2 (the virus that causes COVID disease) used established molecular techniques such as reverse transcription polymerase reaction (RT-PCR). These techniques detect and identify organisms by amplifying their genetic material millions of times. Running the tests however requires trained technicians and expensive equipment.

As the pandemic became more severe, other ways to test for the virus had to be developed. Substances and compounds needed to effectively run diagnostic tests were in short supply and many countries did not have the kinds of sophisticated laboratories needed for the existing tests. Low- and middle-income countries like those throughout the African continent had limited finances too and not enough trained specialists to handle the demand.

> **SHERLOCK and DETECTR are two innovative CRISPR-based kits used for the detection of SARS-CoV-2.**

Isothermal amplification techniques helped to address the need. This is a simple process which rapidly and effectively amplifies DNA and RNA (genetic material) at constant temperature.

Immunological assays also helped. These tests can be used on-site or in the lab and are able to detect specific molecules such as antibodies and antigens. Antibodies are generated in a person's body when a foreign molecule (antigen) invades the body.

These cost-effective tests provide rapid results and can be used on a big scale even where resources are scarce. The major challenge of these tests is that they are less accurate. Unlike molecular tests, which amplify the genetic material of the virus, immunological assays do not amplify their protein signal. This makes them less sensitive. The risk is high that an infected person might incorrectly be told that they don't have the virus.

The global diagnostic community realised it was time to look at methods that were as accurate as conventional molecular tests but could be used outside laboratories and on a large scale.

Big Strides

Scientists needed a new generation of rapid, accurate, accessible and affordable diagnostic tests. The National Institutes of Health in the US set up the Rapid Acceleration of Diagnostics programme (RADx) in 2020 to fund innovative point-of-care and home-based tests and to speed up the development, validation and commercialisation of these tests.

One particularly interesting change in this space is the use of CRISPR. The technology was previously known for its use in gene editing. But now it has revolutionised diagnostics with the launch of SHERLOCK and DETECTR, two innovative CRISPR-based kits used for the detection of SARS-CoV-2. These are particularly sensitive and specific and provide a visual colour readout using a commercially available paper dipstick, making them suitable for use as a point-of-care test.

The versatility of these techniques enables researches to apply the same principles to the detection of other infectious diseases too.

There have also been advances in using biosensors, nanotechnology, smartphone-based tests and wearable technologies for diagnostics.

Overall, in the past three years, the focus of disease testing has moved from simple detecting and understanding to incorporating speed, efficiency and portability of the tests.

Problems Remain

While there is a lot to celebrate in the diagnostic space, problems remain. There are barriers in developing and disseminating tests, particularly in poorer countries. Fairer access to quality testing and improved data sharing between countries is needed to eliminate the inequity in diagnostics.

The lack of resources to deliver a robust regulatory system in low- and middle-income countries also poses a serious challenge. Companies have less incentive to develop and commercialise products where there is weak regulation. Thus countries still depend on tests that are manufactured elsewhere.

As the world moves out of its pandemic response phase, it is likely that investment in diagnostics will fall. With a reduced need for tests, the economic return of investing in developing tests will diminish.

This is unfortunate as there are still so many healthcare challenges worldwide and unless disease surveillance is proactive, it won't be possible to predict where the next pandemic might emerge from. The momentum created by the COVID pandemic offers an opportunity and should be used to build on the things that worked well in the diagnostic industry and to improve on the things that didn't.

Print Citations

CMS: Loots, Angelika. "COVID Testing Led to New Techniques of Disease Diagnosis: Progress Mustn't Stop Now." In *The Reference Shelf: Gene Editing & Genetic Engineering,* edited by Micah L. Issitt, 72–74. Amenia, NY: Grey House Publishing, 2023.

MLA: Loots, Angelika. "COVID Testing Led to New Techniques of Disease Diagnosis: Progress Mustn't Stop Now." *The Reference Shelf: Gene Editing & Genetic Engineering,* edited by Micah L. Issitt, Grey House Publishing, 2023, pp. 72–74.

APA: Loots, A. (2023). COVID testing led to new techniques of disease diagnosis: Progress mustn't stop now. In Micah L. Issitt (Ed.), *The reference shelf: Gene editing & genetic engineering* (pp. 72–74). Amenia, NY: Grey House Publishing. (Original work published 2023)

Forget Designer Babies. Here's How CRISPR Is Really Changing Lives

Antonio Regalado
MIT Technology Review, March 7, 2023

Forget about He Jiankui, the Chinese scientist who created gene-edited babies. Instead, when you think about gene editing you should think of Victoria Gray, the African-American woman who says she's been cured of her sickle-cell disease symptoms.

This week in London, scientists are gathering for the Third International Summit on Human Genome Editing. It's gene editing's big event, where researchers get to awe the audience with their new ability to modify DNA—and ethicists get to worry about what it all means.

The event got underway Monday with a look back at what organizers called the technology's "misuse" in China to create designer babies in 2018. That was certainly an ethical dumpster fire and raised profound questions about whether we should meddle in evolution.

But the designer-baby debate is a distraction from the real story of how gene editing is changing people's lives, through treatments used on adults with serious diseases.

In fact, there are now more than 50 experimental studies underway that use gene editing in human volunteers to treat everything from cancer to HIV and blood diseases, according to a tally shared with MIT Technology Review by David Liu, a gene-editing specialist at Harvard University.

Most of these studies—about 40 of them—involve CRISPR, the most versatile of the gene-editing methods, which was developed only 10 years ago.

The approach was designed to prevent mitochondrial disease, but new evidence shows it might not work as planned.

That is where Gray comes in. She was one of the first patients treated using a CRISPR procedure, in 2019, and when she addressed the group in London, her story left the room in tears.

"I stand here before you today as proof miracles still happen," Gray said of her battle with the disease, in which misshapen blood cells that don't carry enough oxygen can cause severe pain and anemia.

But Gray's case also shows the obstacles facing the first generation of CRISPR treatments, sometimes referred to as "CRISPR 1.0." They will be hugely expensive and tricky to implement, and they could be quickly superseded by a next generation of improved editing drugs.

The company developing Gray's treatment, Vertex Pharmaceuticals, says it's treated more than 75 people in its studies of sickle cell, and a related disease, beta-thalassemia, and that the therapy could be approved for sale in the US within a year. It is widely expected to be the first treatment using CRISPR to go on sale.

Vertex hasn't said what it could cost, but you can expect a price tag in the millions.

A Revelation

Researchers say the technique's march forward to use in medicine has been remarkably fast. "I think CRISPR [has] outpaced every previous genomic therapy technology," says Fyodor Urnov, a researcher at the University of California, Berkeley.

To scientists, CRISPR is a revelation because of how it can snip the genome at specific locations. It's made up of a cutting protein paired with a short gene sequence that acts like GPS, zipping to a predetermined spot in a person's chromosomes.

What's more, it's trivially easy to change that GPS sequence, says Jennifer Doudna, the Berkeley biochemist who shared a Nobel for inventing the method. "CRISPR is a technology that enables changes to DNA that are programmed," she reminded the audience at the summit.

Along with Vertex, a wave of biotech companies, like Intellia, Beam Therapeutics, and Editas Medicine, are hoping they can use this technology to develop successful treatments. Many of them are running the trials on Liu's list. But not all of these trials will be successful.

For instance, in January the San Francisco biotech Graphite Bio had to stop its own tests of a gene-editing treatment for sickle-cell after its first patient's blood cell counts dropped dangerously. The problem was caused by the treatment itself. Graphite's stock has plunged more than 90%, and now the firm's future is in question.

The trick facing all these efforts remains getting CRISPR where it needs to go in the body. That's not easy. In Gray's case, doctors removed bone marrow cells and edited them in the lab. But before they were put back in her body, she underwent punishing chemotherapy to kill off her remaining bone marrow in order to make room for the new cells.

In essence, the Vertex treatment requires a bone marrow transplant. That is an ordeal in itself, and not every patient will be ready for it. Vertex thinks the treatment will be suitable for "severe" cases, a market it estimates includes 32,000 people in Europe and the US.

Even then, patients won't get the treatments if insurers and governments balk at paying. It's a real risk. For instance, a different gene therapy for beta-thalassemia, developed by Bluebird Bio, was pulled out of the European market after governments there refused to pay the $1.8 million price.

CRISPR 2.0

The first generation of CRISPR treatments are also limited in another way. Most use the tool to damage DNA, essentially shutting off genes—a process famously described as "genome vandalism" by Harvard biologist George Church.

Treatments that attempt to break genes include one designed to try to zap HIV. Another is the one Gray got. By breaking a specific bit of DNA, her treatment unlocks a second version of the hemoglobin gene that people normally use only as babies. Since hemoglobin is the errant protein in sickle-cell, booting up another copy solves the problem.

According to Liu's analysis, two-thirds of current studies aim at "disrupting" genes in this way.

Liu's lab is working on next-generation gene-editing approaches. These tools also employ the CRISPR protein, but it's engineered not to cut the DNA helix, but instead to deftly swap individual genetic letters or make larger edits. These are known as "base editors."

According to Lluís Montoliu, a gene scientist at Spain's National Center for Biotechnology, these new versions of CRISPR have "lower risk and better performance," although delivering them "to the right target cell in the body" remains difficult.

At his lab, Montoliu is using base editors to cure mice of albinism, in some cases from birth. It's a step, he says, toward a treatment newborn humans could receive, although not to change their skin color. Instead, he dreams of putting Liu's molecules in their eyes to correct severe vision problems that albinism can cause.

So far, though, the albinism project is not a commercial venture. And that points to one of the biggest limits on CRISPR's impact now and in the foreseeable future. Nearly all CRISPR trials underway aim at either cancer or sickle-cell disease, with multiple companies chasing the exact same problems.

According to Urnov, this means thousands of other inherited diseases that could be treated with CRISPR are just being ignored. "This is near-entirely due to the fact that most of them are too rare to be a viable commercial opportunity," he says.

At the London meeting, however, Urnov will be presenting his ideas on how treatments could be tested even for ultra-rare diseases, including some genetic conditions so unusual they affect just one person.

That's not a commercial opportunity, but because of how CRISPR can be programmed to go anywhere in the genome, it's scientifically possible. Now that gene editing has had its first successes, Urnov says, there's an "urgent need" to open a "path to the clinic for all."

> **Now that gene editing has had its first successes, Urnov says, there's an 'urgent need' to open a 'path to the clinic for all.'**

Print Citations

CMS: Regalado, Antonio. "Forget Designer Babies. Here's How CRISPR Is Really Changing Lives." In *The Reference Shelf: Gene Editing & Genetic Engineering,* edited by Micah L. Issitt, 75–78. Amenia, NY: Grey House Publishing, 2023.

MLA: Regalado, Antonio. "Forget Designer Babies. Here's How CRISPR Is Really Changing Lives." *The Reference Shelf: Gene Editing & Genetic Engineering,* edited by Micah L. Issitt, Grey House Publishing, 2023, pp. 75–78.

APA: Regalado, A. (2023). Forget designer babies. Here's how CRISPR is really changing lives. In Micah L. Issitt (Ed.), *The reference shelf: Gene editing & genetic engineering* (pp. 75–78). Amenia, NY: Grey House Publishing. (Original work published 2023)

4
Controversial Technology

Image by The He Lab, CC BY SA-3.0, via Wikipedia.

Chinese scientist He Jiankui, pictured above in 2018, became widely known that same year after he announced that he had created the first human genetically edited babies. The announcement led to ethical and legal controversies, ultimately resulting in He's imprisonment and widespread international condemnation.

Genetic Modification Controversies and Debates

The first controversies surrounding genetic manipulation were focused on the creation of genetically modified or "GM foods." Concern over food safety and security first became prevalent in the early 1900s, when prepackaged foods began to replace fresh foods across America and consumers had a more difficult time determining what they were consuming. These same kinds of fears resurged with the advent of the first GM foods in the 1990s. Genetically modified crops have modified genes that give the crops properties not present in the original species. Consumers and agriculture and food activists have frequently expressed concern about how the use of genetic modification may affect health over the long term and some have objected to the way that GM agriculture leads to mass monoculturing of crops, which has many negative environmental repercussions.

The very first GM crop on the general market was the "Flavr Savr" tomato, a tomato containing two added genes that increased resistance to fungus and gave the tomato an improved shelf life. The Flavr Savr was introduced to stores in 1994 after having been developed over several years by a California-based biotechnology company. While the product proved to be safe after numerous rounds of testing both in the United States and the UK, safety concerns initially limited sales and led to a movement to have GM products clearly marketed when sold to vendors.[1]

Among the concerns expressed by anti-GM foods activists included the potential for GM crops to replace all existing versions of nonmodified crops, thus reducing human choice and eliminating "species." Other critics were concerned primarily about how GM crops would impact the environment. By far the most pressing concern was that GM crops would harbor hidden health threats that would manifest later. Numerous national studies, including a 2004 study by the National Academies of Sciences, concluded that humans have not been harmed by consuming GM crops.[2] Despite this, there are many Americans who feel that GM foods are unnatural or who feel a sense of revulsion to the consumption of this kind of food whether or not there are any legitimate health concerns.

The Cloning Controversy

The next major controversy surrounding genetic manipulation was the cloning controversy that exploded around the world after the creation of the world's first cloned animal, "Dolly" the sheep. Dolly was created by scientists working at the Roslin Institute of the University of Edinburgh, and she was cloned from a cell taken from the mammary gland of a Finn Dorset sheep, combined with an egg taken from a Scottish Blackface sheep. The clone was born from a Scottish Blackface surrogate in 1996. The sheep died at age six after contracting arthritis and a ovine virus, though neither were related to her having been a cloned animal.[3]

While studies indicating that consuming cloned animals poses no significant health-care risk, the predominant objection to animal cloning has been that the cloning and harvesting process involves more suffering than standard methods of breeding animals and this is because cloning animals involves surgeries to obtain eggs and cells, then to transfer embryos, and, in many cases, to deliver developed offspring via caesarean section. In addition, many cloned births fail and there is a higher degree of pregnancy losses, sickness, and death among young animals. In addition, some ethicists have raised concerns that cloning animals could increase inbreeding and genetic homogeneity among animal populations, which tend to have a deleterious effect over time. In addition, despite the fact that there is no evidence of any health risk associated with the consumption of meat or animal products from cloned animals, many Americans are still wary of consuming cloned animals products and believe that there are potential side effects that have not yet been discovered or revealed. As with other forms of genetic modification, some critics also believe that the cloning of animals is unnatural or violates some laws of nature or spiritual dictates.[4]

While the cloning of all animals has generated controversies, the introduction of the first cloned mammal, the sheep Dolly, also intensified concern about the possibility of cloning humans. This possibility has generated intense controversy for a number of reasons. First, because more than 90 percent of cloning attempts end in failure, the process of cloning a human would likely involve the loss of numerous fetuses and batches of fetal tissue and some Americans consider this unethical or immoral. Further, studies have shown that cloned individuals have a higher risk for developing genetic disorders, cancer, and of having a reduced life expectancy and critics have therefore argued that producing humans more likely to suffer from these conditions is also immoral and unethical.[5]

Further, the broader criticisms of genetic manipulation and the cloning of non-human animals, such as the perception that cloning violates natural laws, is intensified when it comes to the cloning of humans. Many critics have suggested that human cloning is unnatural and a violation of spiritual principles and that scientists who seek to clone humans are essentially playing God. In 2002, a French cult leader named Brigitte Boisselier claimed that the first human clone, a baby named "Eve" had been created outside of the United States. Boisselier was a representative of the Raëlian religious movement, which believes that humanity was created by an alien species that also inspired myths about angels. No evidence of the alleged baby's existence was ever provided and many believed that the claim was an elaborate hoax. Nevertheless, the controversy surrounding the alleged first human clone led to an antihuman-cloning movement in the United States and in many other countries. While the United States did not ban human cloning outright, the United Nations adopted a resolution calling for the international prohibition of human cloning.[6]

The Gene-Editing Controversy

In the 2000s, gene therapy and genetic medicine transformed from a futuristic potential into an immediate reality with scientists conquering many of the barriers that once prevented the genetic modification of humans to eliminate or address disease. While still an uncommon practice, the discovery of new tools, like the use of viral vectors for introducing genes into the body, and the gene-editing tool Clustered Regularly Interspaced Short Palindromic Repeats (CRISPR), made genetic medicine an important part of modern medical development. As genetic medicine developed, many of the same criticisms were aimed at this emerging field, including the perception that altering or interfering with genetic expression is an unnatural practice and/or a violation of natural or spiritual laws. In addition, critics raised concerns about the ways in which genetic manipulation could change the human species.

In 2018, biophysicist He Jiankui announced that he had, for the first time, successfully altered the deoxyribonucleic acid (DNA) of three human embryos, using the CRISPR gene-editing technology, and that these genetically altered embryos, called "CRISPR babies" in the news, were later implanted and birthed by two surrogate women. Dr. He was arrested, convicted, and sentenced to three years imprisonment for violating Chinese laws regarding genetic manipulation, but he defended his actions claiming that the genetic editing of fetuses could help eliminate conditions like "Down syndrome," which results from an extra copy of a chromosome, as well as other genetic conditions. As news of the CRISPR babies spread around the world, a number of countries passed laws prohibiting the genetic alteration of human fetuses or strengthening existing laws.[7]

Unlike human cloning, which raises concerns about cruelty and introducing genetic dysfunction to developing humans, what are the main objections to gene editing fetuses and embryos? If gene editing can safely eliminate potentially deadly diseases like sickle-cell anemia or developmental conditions like Down syndrome, why would humanity avoid utilizing this kind of technology? The controversy surrounding the CRISPR babies spread throughout news outlets around the world and thereby provided important information about the kinds of concerns that people around the world have about gene editing. Writing in the *Canadian Medical Association Journal*, researchers Bartha Maria Knoppers and Erika Kleiderman noted that ethical concerns about gene editing fell into four basic categories:[8]

First, many critics are concerned about the unknown risks to the health and welfare of children from utilizing genetic alteration. Fears of this kind come in many different forms but tend to reflect the idea that this technology is too new and untested to know if genetically altered humans would eventually develop health problems, similarly to how cloned organisms can suffer from a higher potential for genetic abnormalities and disease. Generalized fear or skepticism about scientific development plays a prominent role in this type of criticism and critics may cite past instances in which allegedly safe practices later proved to lead to health concerns.

Second, some critics are concerned that the development of gene editing is occurring before appropriate regulations can be put into place and laws developed to ensure that this technology is not misused or abused by unethical individuals. This form of criticism also draws heavily from a generalized mistrust of scientists, lack of scientific knowledge, and concern about the potential as-yet-unforeseen risks and threats that might arise through genetic manipulation.

Third, some critics believe that rushing forward with gene-editing technology before humanity has adjusted to the possibilities of this kind of genetic medicine will have a "chilling" effect on research. This criticism comes largely from those who are supportive about the idea of using genetic editing to address gene-linked conditions and diseases, but who fear that pushing the technology forward too fast will inspire fear among the public and lead to more restrictive laws that will also prohibit or set back research into the potential benefits of this emerging technology. Critics argued that He's efforts to gene edit human embryos was a step too far for public comfort and that this could ultimately set back global research on the topic.

Finally, some critics are concerned that gene editing is a step towards a new form of "eugenics," a pseudoscience that first became popular in the late 1800s and involved selectively breeding humans to eliminate "undesirable" traits. In the early 1900s, eugenics was basically a scientific-sounding form of white supremacy and white supremacist politicians used the alleged goal of "improving" the species as an excuse to try and limit the reproduction of people of color and other people with allegedly "defects," like blindness, deafness, and many other mental health conditions. As gene editing has evolved from futuristic fantasy to reality, many have raised concerns about how societies might use this technology to eliminate human variety and diversity seen as dysfunctional.

For example, deafness is considered by some to be an "impairment" or a "defect," but many Deaf people do not see themselves as impaired or dysfunctional, but simply argue that they are a different type of human, one that utilizes different methods of communication. The idea that gene editing could eliminate Deafness therefore appears, to some critics, as the elimination not of a "defect" but of a kind of human. Similarly, there are activists who argue that the autism spectrum is not a kind of mental health defect, but simply an example of "neurodiversity," of brains that function differently but are not actually "diseased" or dysfunctional, but simply different. Some critics of gene editing have argued that eliminating these naturally occurring variants of the human species might be seen by some as a positive development, but would actually mean reducing and eliminating some aspects of human diversity.

Controversy and Fear

Much of the controversy surrounding genetic manipulation, whether in the form of GM foods, or cloned animals, or genetically altered fetuses, represents potential fears rather than issues that have been fully realized. Fears about the health concerns of GM foods, for instance, are not based on actual incidents or evidence of any specific health risks but rather on fear of the unknown mingled with fear of corporate and political neglect and willingness to prioritize profit over public safety.

Other issues, like the ethical and moral questions surrounding the use of gene editing to homogenize human bodies and minds delves into deeper territory and raises subjective questions about the nature of humanity and human variation. When it comes to issues like human cloning and the prenatal genetic manipulation, the debate over these technologies have forced people to confront questions about what is "natural" in human life.

Similar questions have been raised over another controversial activity utilizing genetic manipulation, which is the effort to re-create extinct species. While some feel that this effort could enable humanity to replace animals wiped out by human overconsumption and destruction, the "de-extinction" process raises deep questions about the degree to which humanity can or perhaps should interfere or manipulate the processes of speciation and extinction. This recent avenue of genetic science, long an imagined possibility in science fiction, again reflects some of the deeper questions surrounding genetic science in general, which reflect different concepts about nature, human nature, and humanity's role in the planet's biosphere and as a natural species.

Works Used

Bruening, G., and J. M. Lyons. "The Case of the FLAVR SAVR Tomato." *California Agriculture* 54, no. 4, July 1, 2000.

Cohen, Jon. "As Creator of 'CRISPR Babies' Nears Release from Prison, Where Does Embryo Editing Stand?" *Science*. Mar. 21, 2022.

Hinrichs, Katrin. "Controversies About Cloning of Domestic Animals." *Merck Manual*. Nov. 2022. www.merckvetmanual.com/management-and-nutrition/cloning-of-domestic-animals/controversies-about-cloning-of-domestic-animals. Accessed Aug. 2023.

Knoppers, Bartha Maria, and Erika Kleiderman. "'CRISPR Babies': What Does This Mean for Science and Canada?" *Canadian Medical Association Journal*. Jan. 28, 2019.

Langlois, Adèle. "The Global Governance of Human Cloning: The Case of UNES-CO." *Humanities and Social Sciences Communications*. No. 3, 2017.

"The Life of Dolly." *Roslin Institute*. Center for Regenerative Medicine. dolly.roslin.ed.ac.uk/facts/the-life-of-dolly/index.html. Accessed Aug. 2023.

Murayama, Satomi Angelika. "Op-ed: The Dangers of Cloning." *Berkeley*. May 11, 2020. funginstitute.berkeley.edu/news/op-ed-the-dangers-of-cloning/. Accessed Aug. 2023.

National Research Council. *Safety of Genetically Engineered Foods: Approaches to Assessing Unintended Health Effects*. Washington, DC: National Academies of Sciences, 2004.

Notes

1. Bruening and Lyons, "The Case of the FLAVR SAVR Tomato."
2. National Research Council, *Safety of Genetically Engineered Foods: Approaches to Assessing Unintended Health Effects.*"
3. "The Life of Dolly," *Roslin Institute.*
4. Hinrichs, "Controversies About Cloning of Domestic Animals."
5. Murayama, "Op-ed: The Dangers of Cloning."
6. Langlois, "The Global Governance of Human Cloning: The Case of UNES-CO."
7. Cohen, "As Creator of 'CRISPR Babies' Nears Release from Prison, Where Does Embryo Editing Stand?"
8. Knoppers and Kleiderman, "CRISPR Babies: What Does This Mean for Science and Canada?"

Decoding the CRISPR-Baby Stories

By J. Benjamin Hurlbut
MIT Technology Review, February 24, 2021

The conventional story of CRISPR genome editing is one of heroic power and promise with an element of peril. That peril became personified when *MIT Technology Review*'s Antonio Regalado revealed in November 2018 that a young Chinese scientist named He Jiankui was using CRISPR to engineer human embryos. At least three of them became living children. The "CRISPR babies" episode is now an obligatory chapter in any telling of the gene-editing story. When Jennifer Doudna and Emmanuelle Charpentier were awarded the Nobel Prize last year for their invention of CRISPR, virtually every news story also mentioned He. In this century's grandest story of heroic science, he plays the villain.

Storytelling matters. It shapes not only how the past is remembered, but how the future unfolds.

He Jiankui's plans were shaped by stories about how science progresses and how heroes are made. One such moment came in a small, closed-door meeting hosted by Doudna at the University of California, Berkeley, in January 2017, to which He was invited. There a senior scientist from an elite American university observed, "Many major breakthroughs are driven by one or a couple of scientists ... by cowboy science."

I too was at that meeting in January 2017, where I met He for the first time. We exchanged notes periodically in the months that followed, but the next time I saw him was at the International Summit on Genome Editing in Hong Kong in 2018, two days after Regalado had forced him to go public before he planned. After the summit, He disappeared from view: he was being held by Chinese authorities in a guest house on his university's campus.

A month later, he called me, wanting to tell his story. He gave me a detailed history of the CRISPR-babies episode, explaining what motivated his project and the network of people—scientists, entrepreneurs, venture capitalists, and government officials—who supported it. The 2017 Berkeley meeting turned out to have been pivotal, especially the "cowboy science."

After the 2017 meeting, He started reading biographies of scientific risk-takers who were ultimately hailed as heroes, from Edward Jenner, creator of the first vaccine, to Robert Edwards, pioneer of in vitro fertilization (IVF). In January 2019, he wrote to government investigators: "I firmly believe that what I am doing is to promote the progress of human civilization. History will stand on my side."

Looking back at my notes from the 2017 meeting, I discovered that He had remembered only the first half of that provocative statement. It continued: "What's going on right now is cowboy science … but that doesn't mean that's the best way to proceed … we should take a lesson from our history and do better the next time around."

Learning from History?

Kevin Davies's *Editing Humanity* follows a circuitous path through the remarkably diverse experiments and laboratories where the CRISPR puzzle was pieced together. The story of discovery is gripping, not least because Davies, a geneticist turned editor and writer, skillfully weaves together a wealth of detail in a page-turning narrative. The book gives a textured picture of the intersection of academic science with the business of biotechnology, exploring the enormous competition, conflict, and capital that have surrounded CRISPR's commercialization.

However, Davies's book is heavy on the business of gene editing, light on the humanity. The narrative emphasizes the arenas of scientific discovery and technological innovation as though they alone are where the future is made.

Humanity first appears as something more than an object of gene editing in the last line of the book: "CRISPR is moving faster than society can keep up. To where is up to all of us." Yet most of us are missing from the story. Admittedly, the book's focus is the gene editors and their tools. But for readers already primed to see science as the driver of progress, and society as recalcitrant and retrograde until it eventually "catches up," this telling reinforces that consequential myth.

Walter Isaacson's *The Code Breaker* cleaves even more closely to scientific laboratories, following the personalities behind the making of CRISPR. The main protagonist of his sprawling book is Doudna, but it also profiles the many other figures, from graduate students to Nobel laureates, whose work intersected with hers. In always admiring and sometimes loving detail, Isaacson narrates the excitement of discovery, the heat of competition, and the rise of scientific celebrity—and, in He's case, infamy. It is a fascinating story of rivalry and even pettiness, albeit with huge stakes in the form of prizes, patents, profits, and prestige.

Yet for all its detail, the book tells a narrow story. It is a conventional celebration of discovery and invention that sometimes slides into rather breathless celebrity profile (and gossip). Apart from some chapters of Isaacson's own rather superficial ruminations on "ethics," his storytelling rehearses clichés more than it invites reflection and learning. Even the portraits of the people feel distorted by his flattering lens.

The one exception is He, who gets a few chapters as an unwelcome interloper. Isaacson makes little effort to understand his origins and motivations. He is a nobody with a "smooth personality and a thirst for fame" who attempts to force his way into an elite club where he has no business being. Disaster ensues.

He's story ends with a "fair trial" and a prison sentence. Here Isaacson parrots a state media report, unwittingly playing propagandist. The official Chinese story was crafted to conclude the He affair and align Chinese science with the responsible rather than the rogue.

> **Science-centric storytelling implies that science sits outside of society, that it deals primarily with pure arenas of nature and knowledge. But that is a false narrative.**

Authorizing Narratives

These stories of heroic science take for granted what makes a hero—and a villain. Davies's account is considerably more careful and nuanced, but it too shifts to casting stones before seeking to understand the sources of failure—where He's project came from, how a person trained at elite American universities could have believed he would be valorized, not condemned, and how he could get so far without realizing how deep a hole he had dug for himself.

My overwhelming sense from my interviews with He is that far from "going rogue," he was trying to win a race. His failure lay not in refusing to listen to his scientific elders, but in listening too intently, accepting their encouragement and absorbing things said in the inner spaces of science about where genome editing (and humanity) are headed. Things like: CRISPR will save humanity from the burden of disease and infirmity. Scientific progress will prevail as it has always done when creative and courageous pioneers push boundaries. Genome editing of the germline—embryos, eggs, or sperm that will pass changes down to future generations—is inevitable; the only question is who, when, and where.

He heard—and believed in—the messianic promise of the power to edit. As Davies writes, "If fixing a single letter in the genetic code of a fellow human being isn't the coveted chalice of salvation, I don't know what is."

Indeed, as even Isaacson notes, the National Academies had sent similar signals, leaving the door open to germline engineering for "serious diseases or conditions." He Jiankui was roundly criticized for making an edit that was "medically unnecessary"—a genetic change he hoped would make babies genetically resistant to HIV. There are, the critics argued, easier and safer ways to avoid transmitting the virus. But he believed that the terrible stigma in China against HIV-positive people made it a justified target. And the Academies left room for that call: "It is important to note that such concepts as 'reasonable alternatives' and 'serious disease or condition' … are necessarily vague. Different societies will interpret these concepts in the context of their diverse historical, cultural, and social characteristics."

He understood this as an authorization. These are the true origins of his grotesque experiment. The picture of He, and the scientific community he was embedded in, is a rather more ambiguous one than the virtuous science of Isaacson's telling. Or, rather, it's a more human one, in which knowledge and technical acumen aren't necessarily accompanied by wisdom and may instead be colored by ambition, greed, and myopia. Isaacson does the scientists a disservice by presenting them as the makers of the future rather than as people confronting the awesome power of the tools they have created, attempting (and, often, failing) to temper promises of progress with the humility to recognize that they are out of their depth.

Another cost of science-centric storytelling is the way it implies that science sits outside of society, that it deals primarily with the pure arenas of nature and knowledge. But that is a false narrative. For instance, the commercial business of IVF is a crucial part of the story, and yet it receives remarkably little attention in Davies's and Isaacson's accounts. In this regard, their books reflect a deficit in the genome-editing debates. Scientific authorities have tended to proceed as though the world is as governable as a laboratory bench, and as if anyone who thinks rationally thinks like them.

Humanity's Stories

These science-centric stories sideline the people in whose name the research is done. Eben Kirksey's *The Mutant Project* brings those people into the picture. His book, too, is a tour of the actors at the frontiers of genome editing, but for him those actors also include patients, activists, artists, and scholars who engage with disability and disease as lived experiences and not merely as DNA molecules. In Kirksey's book, issues of justice are entangled with the way stories are told about how bodies should be—and not be. This wrests questions of progress from the grip of science and technology.

Like Davies, Kirksey uses the He affair to frame his story. A skilled anthropologist, he is at his best when drawing out people's own stories about what is at stake for them. Some of the most remarkable interviews in the book are with the patients from He Jiankui's trial, including an HIV-positive medical professional who became more deeply committed to He's project after he was fired from his job because his HIV status was discovered.

Kirksey's attention to human beings as more than engineerable bodies, and to the desires that drive the imperative to edit, invites us to recognize the extraordinary peril of reaching into the gene-editing tool kit for salvation.

That peril is too often obscured by hastily spun stories of progress. On the final morning of the genome-editing summit in Hong Kong, less than 24 hours after He had presented his CRISPR-babies experiment, the conference organizing committee issued a statement simultaneously rebuking him and laying a pathway for those who would follow in his footsteps. Behind the statement was a story: one in which technology is racing ahead, and society needs to just accept it—and affirm it. A member of that committee told Kirksey why they had rushed to judgment: "The first person who puts it on paper wins."

So far, the CRISPR story has been about racing to be the first to write—not just scientific papers, but the nucleotides of the genome and rules for the human future. The rush to write—and win—the future leaves little room for learning from patterns of the past. Stories of technological futures, thrilling though they may be, substitute a thin narrative of progress for the richness and fragility of the human story.

We need to listen to more and better storytellers. Our common future depends upon it.

Print Citations

CMS: Hurlbut, J. Benjamin. "Decoding the CRISPR-Baby Stories." In *The Reference Shelf: Gene Editing & Genetic Engineering,* edited by Micah L. Issitt, 87–91. Amenia, NY: Grey House Publishing, 2023.

MLA: Hurlbut, J. Benjamin. "Decoding the CRISPR-Baby Stories." *The Reference Shelf: Gene Editing & Genetic Engineering,* edited by Micah L. Issitt, Grey House Publishing, 2023, pp. 87–91.

APA: Hurlbut, J. B. (2023). Decoding the CRISPR-baby stories. In Micah L. Issitt (Ed.), *The reference shelf: Gene editing & genetic engineering* (pp. 87–91). Amenia, NY: Grey House Publishing. (Original work published 2021)

It's Official: No More CRISPR Babies—for Now

By Grace Browne
Wired, March 17, 2023

Last week in London, a small group of protestors braved it out in the rain in front of the Francis Crick Institute, where the Third International Summit on Human Genome Editing was taking place. The sparse congregation, from the group Stop Designer Babies, brandished signs urging "Never Again to Eugenics" and "NO HGM"(no human genetic modification). The group campaigns against what it sees as the scientific community's lurch towards using gene editing for biological enhancement—to tweak genomes to give, say, higher intelligence or blue eyes. If this came to pass, it would be a slippery slope towards eugenics, the group argues.

Three days later, at the close of the summit, it seems the group's wishes may have been partially granted—at least for the time being.

After several days of experts chewing on the scientific, ethical, and governance issues associated with human genome editing, the summit's organizing committee put out its closing statement. Heritable human genome editing—editing embryos that are then implanted to establish a pregnancy, which can pass on their edited DNA—"remains unacceptable at this time," the committee concluded. "Public discussions and policy debates continue and are important for resolving whether this technology should be used."

The use of the word "whether" in that last sentence was carefully selected and carries a lot of weight, says Françoise Baylis, a bioethicist who was on the organizing committee. Crucially, the word isn't "how"—"that, I think, is a clear signal to say the debate's open," she says.

This marks a shift in attitude since the close of the last summit, in 2018, during which Chinese scientist He Jiankui dropped a bombshell: He revealed that he had previously used Crispr to edit human embryos, resulting in the birth of three Crispr-edited babies—much to the horror of the summit's attendees and the rest of the world. In its closing statement, the committee condemned He Jiankui's premature actions, but at the same time it signaled a yellow rather than red light on germline genome editing—meaning, proceed with caution. It recommended setting up a "translational pathway" that could bring the approach to clinical trials in a rigorous, responsible way.

In the intervening half a decade or so, research has confirmed that germline genome editing is still way too risky—and that's before even beginning to grapple with the massive ethical concerns and societal ramifications. And these concerns were only compounded at this year's summit.

> **Heritable human genome editing—editing embryos that are then implanted to establish a pregnancy, which can pass on their edited DNA—"remains unacceptable at this time."**

These include, for example, mosaicism, where genome editing results in some cells getting different edits to others. At the summit, Shoukhrat Mitalipov, a biologist at Oregon Health and Science University, presented findings from his lab that showed that germline genome editing had resulted in unintended—and potentially dangerous—tweaks to the genomes of embryos, which standard DNA-reading tests used to screen embryos before implantation might not pick up. Another scientist, Dagan Wells, a reproductive biologist at the University of Oxford, presented research that looked at how embryos repair breaks in their DNA after having been edited. His work found that about two-fifths of the embryos failed to repair the broken DNA. A child that grows from such an embryo could suffer health problems.

The message was loud and clear: Scientists don't yet know how to safely edit embryos.

To Katie Hasson, associate director of the Center for Genetics and Society, a California nonprofit that advocates for a broad prohibition of heritable genome editing, those few lines in the committee's closing statement were the most important thing to come out of the summit. "I think this is an important step back from the brink."

But figuring out "whether" heritable germline editing will ever be acceptable requires a lot more work. "That conversation about whether we should do it or not needs to be much broader than what we saw at the summit," says Hasson. The world needs to reach broad societal consensus on this question, Baylis says. She worries that that work won't happen. Up until now, these summits have led the discussion on where the field goes, but it's still up in the air whether a fourth summit will ever take place. "I think we haven't yet had the tough conversations that we still need to have," says Baylis.

Print Citations

CMS: Browne, Grace. "It's Official: No More CRISPR Babies—for Now." In *The Reference Shelf: Gene Editing & Genetic Engineering*, edited by Micah L. Issitt, 92–94. Amenia, NY: Grey House Publishing, 2023.

MLA: Browne, Grace. "It's Official: No More CRISPR Babies—for Now." *The Reference Shelf: Gene Editing & Genetic Engineering*, edited by Micah L. Issitt, Grey House Publishing, 2023, pp. 92–94.

APA: Browne, G. (2023). It's official: No more CRISPR babies—for now. In Micah L. Issitt (Ed.), *The reference shelf: Gene editing & genetic engineering* (pp. 92–94). Amenia, NY: Grey House Publishing. (Original work published 2023)

The First Gene-Edited Babies Are Supposedly Alive and Well, Says Guy Who Edited Them

By Tim Newcomb
Popular Mechanics, February 7, 2023

- He Jiankui gene-edited three babies in 2018 and 2019 and claims they are all doing well.
- Jiankui was sent to prison for three years in China for his unethical practices.
- The original goal was to edit genes to offer complete or partial HIV resistance, which may not have worked anyway.

He Jiankui shocked the scientific community in 2018 by announcing his team had used the CRISPR-Cas9 gene-editing tool on twin girls when they were just embryos, resulting in the birth of the world's first genetically modified babies. A third gene-edited child was born a year later.

Now, the disgraced gene-editing scientist, who was imprisoned in China for three years for the unethical practices, tells the *South China Morning Post* that all three children are doing well. "They have a normal, peaceful, and undisturbed life," He says. "This is their wish, and we should respect them. The happiness of the children and their families should come first."

He's original goal was to use gene editing to attempt—many call this a live human experiment—to rewrite the CCR5 gene to create resistance to HIV. He says the genes were edited successfully and believed it gave the babies either complete or partial HIV resistance because of the mutation.

He's claim of success is unsubstantiated. In fact, it remains unclear if the experiment was even effective at all, setting aside for a minute the (huge) ethical implications. According to a 2019 MIT *Technology Review* report, "the team didn't actually reproduce the known mutation. Rather, they created new mutations, which might lead to HIV resistance, but might not."

He tells the *SCMP* he anticipates and worries for the future of the three girls as a traditional father would. He plans to track their medical needs and hopes to raise money to pay for health-related expenses. "After the age of 18," He says, "the children will decide whether to do medical follow-ups for their individual needs. We committed to doing this for their lifetimes."

He, who was released from prison in April 2022, admits his experiments were rushed. "I did it too quickly," he says. Still, the scientist has already set up a Beijing laboratory to work on gene therapies for genetic diseases.

The ethical dilemma that He introduced may just be the start of a long line of questions regarding gene editing. He and his three test subjects won't be the only players in the saga, but they'll always be on the main stage.

> He's claim of success is unsubstantiated. In fact, it remains unclear if the experiment was even effective at all, setting aside for a minute the (huge) ethical implications.

Print Citations

CMS: Newcomb, Tim. "The First Gene-Edited Babies Are Supposedly Alive and Well, Says Guy Who Edited Them." In *The Reference Shelf: Gene Editing & Genetic Engineering*, edited by Micah L. Issitt, 95–96. Amenia, NY: Grey House Publishing, 2023.

MLA: Newcomb, Tim. "The First Gene-Edited Babies Are Supposedly Alive and Well, Says Guy Who Edited Them." *The Reference Shelf: Gene Editing & Genetic Engineering*, edited by Micah L. Issitt, Grey House Publishing, 2023, pp. 95–96.

APA: Newcomb, T. (2023). The first gene-edited babies are supposedly alive and well, says guy who edited them. In Micah L. Issitt (Ed.), *The reference shelf: Gene editing & genetic engineering* (pp. 95–96). Amenia, NY: Grey House Publishing. (Original work published 2023)

Biotechnology Is Creating Ethical Worries and We've Been Here Before

By Giana Gitig
Ars Technica, October 29, 2022

Matthew Cobb is a zoologist and author whose background is in insect genetics and the history of science. Over the past decade or so, as CRISPR was discovered and applied to genetic remodeling, he started to get concerned—afraid, actually—about three potential applications of the technology. He's in good company: Jennifer Doudna, who won the Nobel Prize in Chemistry in 2020 for discovering and harnessing CRISPR, is afraid of the same things. So he decided to delve into these topics, and *As Gods: A Moral History of the Genetic Age* is the result.

Summing Up Fears

The first of his worries is the notion of introducing heritable mutations into the human genome. He Jianqui did this to three human female embryos in China in 2018, so the three girls with the engineered mutations that they will pass on to their kids (if they're allowed to have any) are about four now. Their identities are classified for their protection, but presumably their health is being monitored, and the poor girls have probably already been poked and prodded incessantly by every type of medical specialist there is.

The second is the use of gene drives. These allow a gene to copy itself from one chromosome in a pair to the other so it will be passed on to almost all offspring. If that gene causes infertility, the gene drive spells the extinction of the population that carries it. Gene drives have been proposed as a way to eradicate malaria-bearing mosquitoes, and they have been tested in the lab, but the technology has not been deployed in the wild yet.

Although eliminating malaria seems like an unalloyed good, no one is really sure what would happen to an ecosystem if we get rid of all of the malaria-bearing mosquitoes. (Of course, humans have eliminated or at least severely depleted entire species before—passenger pigeons, bison, eastern elk, wolves—sometimes even on purpose but never with the awareness of the Interconnectedness of All Things that we have now.) Another barrier comes from the fact that deploying this technology hinges on informed consent by the local population, which is difficult when some local languages don't have a word for "gene."

The third concern is focused on gain-of-function studies that create more transmissible or pathogenic viruses in a laboratory. These studies are purportedly done to get a better understanding of what makes viruses more dangerous, so in an ideal world, we could prepare for the eventuality of one occurring naturally. National Institutes of Health-funded gain-of-function studies done in 2011 made the very lethal H5N1 strain of flu more transmissible, leading to a self-imposed research moratorium that ended with more stringent regulations (in some countries). These types of studies obviously have the potential to create bioweapons, and even without nefarious intent, leaks are not impossible. (It is not likely that work of this sort caused the COVID-19 pandemic; evidence suggests that it jumped to humans from wildlife.)

> Humans have eliminated or at least severely depleted entire species before—passenger pigeons, bison, eastern elk, wolves—sometimes even on purpose, but never with the awareness of the Interconnectedness of All Things that we have now.

The title of the resulting book is lifted from Stewart Brand's *Whole Earth Catalog*, in which he wrote: "We are as gods and might as well get good at it." Alas, not all gods are magnanimous. Or even competent, much less good at it.

Calling a Timeout

As a historian of science, Cobb spends much of the book putting his fears in context. One way he does this is by considering how society dealt with the scary, potentially dangerous, and far-reaching advances in genetic manipulation that occurred in the latter half of the 20th century, and then comparing that to how society dealt with the scary, potentially dangerous, and far-reaching advances in nuclear physics that occurred in the former half.

He uses the change in the origin story in the X-Men comics to trace how public fears about science shifted from the atom to the gene. In the 1960s, the X-Men gained their mutations and accompanying powers through radiation exposure; by the 1980s, they were the products of genetic engineering experiments by the long-ago alien Celestials. . . .

The Asilomar conference, held in California in February 1975, is generally held up as a paradigm of self-regulation. At the time, scientists were in the process of establishing recombinant DNA technology—the ability to move genes between organisms and to express any given gene essentially at will in bacteria. It is astonishing that, in the middle of these developments, they decided to pause and debate if and how they should proceed. (This shuffling of genes among species also happens in nature, but they didn't know that yet.) Cobb writes that "no group of scientists, apart from geneticists, has ever voluntarily paused their work because they feared the consequences of what they might discover."

But the Asilomar conference didn't happen because geneticists are more moral than other scientists, Cobb maintains; they were just responding to the fears prevalent at their time. Many of the young researchers who advanced genetic engineering techniques came of age scientifically in the late 1960s, when they were at university protests against the Vietnam War. Between Hiroshima and Nagasaki and Agent Orange, physicists and then chemists watched with horror as the military-industrial complex turned their research into mass death and turned the public against the enterprise of science. These newly minted molecular biologists wanted to ensure that the same thing didn't happen to them, Cobb argues.

Why Are We Worrying?

Cobb further mitigates any pride biologists might feel about this remarkable self-regulation by pointing out that the scientists involved were solely concerned with the potential safety hazards of their work and never discussed any of its moral, ethical, environmental, political, or social impacts. They also didn't invite anyone outside of their immediate field. By keeping the conversation among experts and not addressing society's real fears—like frankenfoods and designer babies—head-on, they allowed those fears to fester, culminating in the type of anti-vaccine, anti-GMO rhetoric we have to contend with today.

Similarly, when CRISPR became a thing, there were experts who urged caution rather than haste. But they also similarly focused only on the potential safety concerns of gene editing and, again, did not deal with any ethical implications. (Despite the exclusive focus on biosecurity at Asilomar, three Soviet delegates who were there went home and started an offensive biological weapons program, giving it the totally innocuous-sounding name Biopreparat. It has since been privatized.)

Once we are squarely oriented on the history leading up to CRISPR, the book goes through the requisite hot-button topics: government regulation of genetic research, the commercialization and patentability of genetic research, the generation of pathogens as bioweapons, genetically modified plant and animal foodstuffs, Impossible Burgers, Monsanto, Russia, China, gene therapy, and the efforts at de-extincting the wooly mammoth. Snarky, self-important footnotes appear periodically. Within each topic, Cobb asserts that there are often lower-tech, less expensive, less ethically fraught but less flashy ways to achieve what genetic editing claims to be trying to achieve. Clean air and water could save more lives than gene therapies, to give one example.

An Inevitably Late Question

Gutenberg couldn't foresee *Mein Kampf* any more than he could foresee *A Wrinkle in Time* or *To Kill a Mockingbird*; the Wright brothers presumably didn't foresee either the Enola Gay or the Blue Angels. But even if they never asked "why" or "what if" about their endeavors, that doesn't mean today's scientists should follow suit. The first genetic engineers *did* foresee the nefarious applications and world-changing consequences of the technology they were introducing, along with the good ones. When the tapes of the Asilomar conference come out of embargo in 2025, the world will judge if they appropriately addressed their reservations.

This book, like so many others on the perils of any technological advancement, asks: Just because we can, does that mean we should? But it is always too late to ask that question. History has shown us over and over again that if we can, we will. We just have to try to proceed responsibly. Whatever that means.

Print Citations

CMS: Gitig, Giana. "Biotechnology Is Creating Ethical Worries and We've Been Here Before." In *The Reference Shelf: Gene Editing & Genetic Engineering,* edited by Micah L. Issitt, 97–100. Amenia, NY: Grey House Publishing, 2023.

MLA: Gitig, Giana. "Biotechnology Is Creating Ethical Worries and We've Been Before." *The Reference Shelf: Gene Editing & Genetic Engineering,* edited by Micah L. Issitt, Grey House Publishing, 2023, pp. 97–100.

APA: Gitig, G. (2023). Biotechnology is creating ethical worries and we've been here before. In Micah L. Issitt (Ed.), *The reference shelf: Gene editing & genetic engineering* (pp. 97–100). Amenia, NY: Grey House Publishing. (Original work published 2022)

Cloning Fact Sheet

National Human Genome Research Institute/NIH, August 15, 2020

The term cloning describes a number of different processes that can be used to produce genetically identical copies of a biological entity. The copied material, which has the same genetic makeup as the original, is referred to as a clone. Researchers have cloned a wide range of biological materials, including genes, cells, tissues and even entire organisms, such as a sheep.

Do Clones Ever Occur Naturally?

Yes. In nature, some plants and single-celled organisms, such as bacteria, produce genetically identical offspring through a process called asexual reproduction. In asexual reproduction, a new individual is generated from a copy of a single cell from the parent organism.

Natural clones, also known as identical twins, occur in humans and other mammals. These twins are produced when a fertilized egg splits, creating two or more embryos that carry almost identical DNA. Identical twins have nearly the same genetic makeup as each other, but they are genetically different from either parent.

What Are the Types of Artificial Cloning?

There are three different types of artificial cloning: gene cloning, reproductive cloning and therapeutic cloning.

Gene cloning produces copies of genes or segments of DNA. Reproductive cloning produces copies of whole animals. Therapeutic cloning produces embryonic stem cells for experiments aimed at creating tissues to replace injured or diseased tissues.

Gene cloning, also known as DNA cloning, is a very different process from reproductive and therapeutic cloning. Reproductive and therapeutic cloning share many of the same techniques, but are done for different purposes.

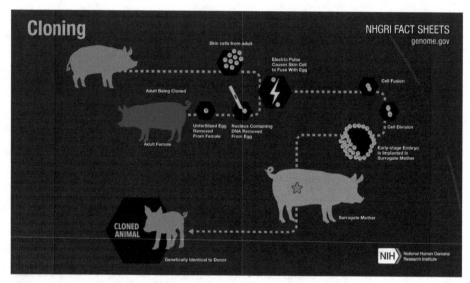

What Sort of Cloning Research Is Going on at NHGRI?

Gene cloning is the most common type of cloning done by researchers at NHGRI. NHGRI researchers have not cloned any mammals and NHGRI does not clone humans.

How Are Genes Clones?

Researchers routinely use cloning techniques to make copies of genes that they wish to study. The procedure consists of inserting a gene from one organism, often referred to as "foreign DNA," into the genetic material of a carrier called a vector. Examples of vectors include bacteria, yeast cells, viruses or plasmids, which are small DNA circles carried by bacteria. After the gene is inserted, the vector is placed in laboratory conditions that prompt it to multiply, resulting in the gene being copied many times over.

How Are Animals Cloned?

In reproductive cloning, researchers remove a mature somatic cell, such as a skin cell, from an animal that they wish to copy. They then transfer the DNA of the donor animal's somatic cell into an egg cell, or oocyte, that has had its own DNA-containing nucleus removed.

Researchers can add the DNA from the somatic cell to the empty egg in two different ways. In the first method, they remove the DNA-containing nucleus of the somatic cell with a needle and inject it into the empty egg. In the second approach, they use an electrical current to fuse the entire somatic cell with the empty egg.

In both processes, the egg is allowed to develop into an early-stage embryo in the test-tube and then is implanted into the womb of an adult female animal.

Ultimately, the adult female gives birth to an animal that has the same genetic make up as the animal that donated the somatic cell. This young animal is referred to as a clone. Reproductive cloning may require the use of a surrogate mother to allow development of the cloned embryo, as was the case for the most famous cloned organism, Dolly the sheep.

What Animals Have Been Cloned?

Over the last 50 years, scientists have conducted cloning experiments in a wide range of animals using a variety of techniques. In 1979, researchers produced the first genetically identical mice by splitting mouse embryos in the test tube and then implanting the resulting embryos into the wombs of adult female mice. Shortly after that, researchers produced the first genetically identical cows, sheep and chickens by transferring the nucleus of a cell taken from an early embryo into an egg that had been emptied of its nucleus.

It was not until 1996, however, that researchers succeeded in cloning the first mammal from a mature (somatic) cell taken from an adult animal. After 276 attempts, Scottish researchers finally produced Dolly, the lamb from the udder cell of a 6-year-old sheep. Two years later, researchers in Japan cloned eight calves from a single cow, but only four survived.

Besides cattle and sheep, other mammals that have been cloned from somatic cells include: cat, deer, dog, horse, mule, ox, rabbit and rat. In addition, a rhesus monkey has been cloned by embryo splitting.

Have Humans Been Cloned?

Despite several highly publicized claims, human cloning still appears to be fiction. There currently is no solid scientific evidence that anyone has cloned human embryos.

In 1998, scientists in South Korea claimed to have successfully cloned a human embryo, but said the experiment was interrupted very early when the clone was just a group of four cells. In 2002, Clonaid, part of a religious group that believes humans were created by extraterrestrials, held a news conference to announce the birth of what it claimed to be the first cloned human, a girl named Eve. However, despite repeated requests by the research community and the news media, Clonaid never provided any evidence to confirm the existence of this clone or the other 12 human clones it purportedly created.

In 2004, a group led by Woo-Suk Hwang of Seoul National University in South Korea published a paper in the journal *Science* in which it claimed to have created a cloned human embryo in a test tube. However, an independent scientific committee later found no proof to support the claim and, in January 2006, *Science* announced that Hwang›s paper had been retracted.

From a technical perspective, cloning humans and other primates is more difficult than in other mammals. One reason is that two proteins essential to cell division, known as spindle proteins, are located very close to the chromosomes in primate eggs. Consequently, removal of the egg's nucleus to make room for the donor nucleus also removes the spindle proteins, interfering with cell division. In other mammals, such as cats, rabbits and mice, the two spindle proteins are spread throughout the egg. So, removal of the egg's nucleus does not result in loss of spindle proteins. In addition, some dyes and the ultraviolet light used to remove the egg's nucleus can damage the primate cell and prevent it from growing.

Do Cloned Animals Always Look Identical?

No. Clones do not always look identical. Although clones share the same genetic material, the environment also plays a big role in how an organism turns out.

For example, the first cat to be cloned, named Cc, is a female calico cat that looks very different from her mother. The explanation for the difference is that the color and pattern of the coats of cats cannot be attributed exclusively to genes. A biological phenomenon involving inactivation of the X chromosome (See sex chromosome) in every cell of the female cat (which has two X chromosomes) determines which coat color genes are switched off and which are switched on. The distribution of X inactivation, which seems to occur randomly, determines the appearance of the cat's coat.

What Are the Potential Applications of Cloned Animals?

Reproductive cloning may enable researchers to make copies of animals with the potential benefits for the fields of medicine and agriculture.

For instance, the same Scottish researchers who cloned Dolly have cloned other sheep that have been genetically modified to produce milk that contains a human protein essential for blood clotting. The hope is that someday this protein can be purified from the milk and given to humans whose blood does not clot properly. Another possible use of cloned animals is for testing new drugs and treatment strategies. The great advantage of using cloned animals for drug testing is that they are all genetically identical, which means their responses to the drugs should be uniform rather than variable as seen in animals with different genetic make-ups.

After consulting with many independent scientists and experts in cloning, the U.S. Food and Drug Administration (FDA) decided in January 2008 that meat and milk from cloned animals, such as cattle, pigs and goats, are as safe as those from non-cloned animals. The FDA action means that researchers are now free to using cloning methods to make copies of animals with desirable agricultural traits, such as high milk production or lean meat. However, because cloning is still very expensive, it will likely take many years until food products from cloned animals actually appear in supermarkets.

Another application is to create clones to build populations of endangered, or possibly even extinct, species of animals. In 2001, researchers produced the first clone of an endangered species: a type of Asian ox known as a guar. Sadly, the baby guar, which had developed inside a surrogate cow mother, died just a few days after its birth. In 2003, another endangered type of ox, called the Banteg, was successfully cloned. Soon after, three African wildcats were cloned using frozen embryos as a source of DNA. Although some experts think cloning can save many species that would otherwise disappear, others argue that cloning produces a population of genetically identical individuals that lack the genetic variability necessary for species survival.

Some people also have expressed interest in having their deceased pets cloned in the hope of getting a similar animal to replace the dead one. But as shown by Cc the cloned cat, a clone may not turn out exactly like the original pet whose DNA was used to make the clone.

> **Researchers have observed some adverse health effects in sheep and other mammals that have been cloned: an increase in birth size and a variety of defects in vital organs, such as the liver, brain, and heart.**

What Are the Potential Drawbacks of Cloning Animals?

Reproductive cloning is a very inefficient technique and most cloned animal embryos cannot develop into healthy individuals. For instance, Dolly was the only clone to be born live out of a total of 277 cloned embryos. This very low efficiency, combined with safety concerns, presents a serious obstacle to the application of reproductive cloning.

Researchers have observed some adverse health effects in sheep and other mammals that have been cloned. These include an increase in birth size and a variety of defects in vital organs, such as the liver, brain and heart. Other consequences include premature aging and problems with the immune system. Another potential problem centers on the relative age of the cloned cell's chromosomes. As cells go through their normal rounds of division, the tips of the chromosomes, called telomeres, shrink. Over time, the telomeres become so short that the cell can no longer divide and, consequently, the cell dies. This is part of the natural aging process that seems to happen in all cell types. As a consequence, clones created from a cell taken from an adult might have chromosomes that are already shorter than normal, which may condemn the clones' cells to a shorter life span. Indeed, Dolly, who was cloned from the cell of a 6-year-old sheep, had chromosomes that were shorter than those of other sheep her age. Dolly died when she was six years old, about half the average sheep's 12-year lifespan.

What Is Therapeutic Cloning?

Therapeutic cloning involves creating a cloned embryo for the sole purpose of producing embryonic stem cells with the same DNA as the donor cell. These stem cells can be used in experiments aimed at understanding disease and developing new treatments for disease. To date, there is no evidence that human embryos have been produced for therapeutic cloning.

The richest source of embryonic stem cells is tissue formed during the first five days after the egg has started to divide. At this stage of development, called the blastocyst, the embryo consists of a cluster of about 100 cells that can become any cell type. Stem cells are harvested from cloned embryos at this stage of development, resulting in destruction of the embryo while it is still in the test tube.

What Are the Potential Applications of Therapeutic Cloning?

Researchers hope to use embryonic stem cells, which have the unique ability to generate virtually all types of cells in an organism, to grow healthy tissues in the laboratory that can be used replace injured or diseased tissues. In addition, it may be possible to learn more about the molecular causes of disease by studying embryonic stem cell lines from cloned embryos derived from the cells of animals or humans with different diseases. Finally, differentiated tissues derived from ES cells are excellent tools to test new therapeutic drugs.

What Are the Potential Drawbacks of Therapeutic Cloning?

Many researchers think it is worthwhile to explore the use of embryonic stem cells as a path for treating human diseases. However, some experts are concerned about the striking similarities between stem cells and cancer cells. Both cell types have the ability to proliferate indefinitely and some studies show that after 60 cycles of cell division, stem cells can accumulate mutations that could lead to cancer. Therefore, the relationship between stem cells and cancer cells needs to be more clearly understood if stem cells are to be used to treat human disease.

What Are Some of the Ethical Issues Related to Cloning?

Gene cloning is a carefully regulated technique that is largely accepted today and used routinely in many labs worldwide. However, both reproductive and therapeutic cloning raise important ethical issues, especially as related to the potential use of these techniques in humans.

Reproductive cloning would present the potential of creating a human that is genetically identical to another person who has previously existed or who still exists. This may conflict with long-standing religious and societal values about human dignity, possibly infringing upon principles of individual freedom, identity and autonomy. However, some argue that reproductive cloning could help sterile couples fulfill their dream of parenthood. Others see human cloning as a way to avoid passing on a deleterious gene that runs in the family without having to undergo embryo screening or embryo selection.

Therapeutic cloning, while offering the potential for treating humans suffering from disease or injury, would require the destruction of human embryos in the test tube. Consequently, opponents argue that using this technique to collect embryonic stem cells is wrong, regardless of whether such cells are used to benefit sick or injured people.

Print Citations

CMS: National Human Genome Research Institute/NIH. "Cloning Fact Sheet." In *The Reference Shelf: Gene Editing & Genetic Engineering,* edited by Micah L. Issitt, 101–107. Amenia, NY: Grey House Publishing, 2023.

MLA: National Human Genome Research Institute/NIH. "Cloning Fact Sheet." *The Reference Shelf: Gene Editing & Genetic Engineering,* edited by Micah L. Issitt, Grey House Publishing, 2023, pp. 101–107.

APA: National Human Genome Research Institute/NIH. (2023). Cloning fact sheet. In Micah L. Issitt (Ed.), *The reference shelf: Gene editing & genetic engineering* (pp. 101–107). Amenia, NY: Grey House Publishing. (Original work published 2020)

5
Popular Genetics

Image by Thomas Quine, CC BY 2.0, via Wikimedia.

De-extinction (by cloning or gene editing and selective breeding) attempts to create an organism that either resembles or is an extinct or endangered species. The Woolly mammoth, picture above in model form at the Royal BC Museum in Victoria, could be a candidate.

Genetic Engineering in Pop Culture

In 1990, author Michael Crichton published the novel *Jurassic Park*, about a fictional theme park populated by dinosaurs genetically engineered by blending recovered dinosaur deoxyribonucleic acid (DNA) with genes from modern amphibians, birds, and reptiles. Due to human error and greed, the dinosaurs escape containment and wreak havoc on the earth.[1] Half monster adventure and half cautionary tale of unrestrained scientific development, *Jurassic Park* became one of the most popular franchises in American history, spawning numerous films and a television series. While humanity's enduring fascination with prehistoric life partially explains the popularity of this series, interest in *Jurassic Park* also reflects a common human fear of, essentially, other humans and how they might misuse and abuse technology to the detriment of other humans.

Over the subsequent thirty years, research suggested that the revival of dinosaurs was likely more fantasy than potential reality, but researchers armed with new tools and techniques were actively attempting to engage in their own, much milder, *Jurassic Park* experiments. Scientists taking part in the "de-extinction" process, as it has been called, seek to revive extinct species like the dodo, the passenger pigeon, and the Carolina parakeet that much more recently disappeared from the earth and did so because of human persecution. These efforts set off a debate among scientists about the ethics and morality of reintroducing species into a world in which they would likely no longer belong. De-extinction is seen by some as an effort to undo humanity's ecological mismanagement, but the question is, can these species survive in the world that humanity has created? For others, the appeal is more immediate and aesthetic, giving humanity a chance to see and interact with species that once existed alongside humans and perhaps even species like the mammoth that predate history.[2] While the creatures targeted for de-extinction are not exactly the frightening prehistoric beasts of Crichton's fiction, the de-extinction effort raises similar concerns and has inspired a debate about human hubris and the unintended and chaotic side effects of messing with the "natural" order of things.[3]

Mucking About with Nature

Jurassic Park wasn't the first book exploring human genetic engineering to become a cultural phenomenon. Arguably, the 1818 novel *Frankenstein* by Mary Shelley, was the first novel in this mold. In *Frankenstein*, the titular doctor, an expert in human anatomy and physiology, learns of a process that can reanimate a deceased corpse. Victor Frankenstein, obsessed with exploring this new scientific reality, ignores the practical issues at hand and, in so doing, creates a homunculus out of place in the world, rejected by humanity, and angry with his creator for having birthed him into a life of loneliness, isolation, and misery. Often cited as the very first "science fiction" book, Shelley's *Frankenstein* explored the fear of how unbridled experimentation with natural forces and properties might backfire. For some, the novel is an exploration of why humanity shouldn't engage in activities that are the domain of "God," while, for others, the novel is a cautionary warning against overzealous interference with nature. However the story is imagined, the basic idea is that the march for progress might eventually lead humanity into arenas of exploration for which humanity is not prepared and might lead to unimagined consequences. This same fear has since played a major role in shaping public attitudes about things like in vitro fertilization (IVF), organ transplantation, genetic medicine, and human cloning.[4]

More closely related to later science-fiction works like *Jurassic Park* and also to the modern era of gene editing and genetic recombination is the 1896 book *The Island of Doctor Moreau*, by English novelist H. G. Wells. The novel tells the story of a shipwrecked man who has the misfortune of ending up on an island occupied by a mad scientist seeking to transform animals into humanistic creatures and who has created a population of "beast men" who are dangerously close to reverting to their "wild" ways. Similar to *Frankenstein*, Wells's *book* explores the theme of unscrupulous scientific experimentation and its potentially disastrous consequences, but the novel also reflects other popular issues in British pop culture of the time. In the late 1800s, many in Britain and in the United States were concerned about "regression" in the human species, a pseudoscientific fear that largely reflected racial and ethnic prejudices, and this pop cultural obsession was also reflected in Wells's writing. Further, the late 1890s was a period of increasing interest in animal welfare and this theme is also present in *Doctor Moreau* as the author reflects on the callous suffering inflicted by the fictional doctor upon the animals that he made into animal-human hybrids. At the same time that the book gained popularity, there was an active movement in Britain to end experimentation on live animals and this broader debate over the morality of animal experimentation found its way into Wells's fiction.[5]

Both *Frankenstein* and *The Island of Doctor Moreau* demonstrate that human fears about interfering in natural processes have long been part of western culture, likely accompanying humanity at every stage. In each generation, new novels, films, and television series exploring this same basic theme appear, often reflecting the intervening advancements in technology. Thus, while *Doctor Moreau* reflected human concern over existing technological processes like the vivisection of animals and the prevailing racially prejudiced belief in human degeneration, more modern versions, like *Jurassic Park*, drew on contemporary developments in science. When *Jurassic Park* hit bookstores, discussions of genetic manipulation were commonplace in part because of the introduction of the first genetically modified (GM) foods like the "Flavr Savr" tomato. This vegetal Frankenstein inspired a national debate about the potential practical or health risks of GM foods, but also reflected the more ancient concern about the potential consequences of humanity's interference in nature or in "spiritual creation," depending on what a person chooses to believe.

Many other storytellers have adopted similar themes, some of which reflect other deep concerns about the push and pull between conservation and advancement. Many of the novels in the modern vein, like the 2017 novel *Borne*, by Jeff Vander-Meer also, like *Jurassic Park*, use genetic manipulation accidents and unintended catastrophes to critique human greed and avarice. In *Jurassic Park*, the desire to earn revenues by showcasing dinosaurs leads scientists to abandon commonsense concerns about the potentially disastrous impact of their experimentation. Likewise, the park's security system is compromised by a greedy employee who shuts down the system to sell stolen genetic material to a rival company in an example of corporate espionage. In further installments of the series, corporations are seen to have ignored the human cost of their unstable experiments with dinosaur life and have engaged in further reckless experimentation in an effort to continue drawing profit. Fiction in this theme reflects concern over the role that greed and corporate indifference to human welfare has played in humanity's past and how economic factors, when applied to science, has the potential to have unintended consequences for humanity.

Shaping Realistic Ideas About Science

Most of the scientists who participate in fields like genetic engineering and genetic medicine are aware of popular concerns and skepticism towards this field of scientific development and also of how the depiction of scientific fields like cloning, gene editing, animal experimentation, and genetic fetal manipulation, have shaped attitudes about genetic science. Scientists involved in genetics have long feared that public misconceptions of genetic science, in part fueled by ill-advised scientific experimentation, could have a "chilling effect" on research, motivating unnecessarily restrictive laws that are meant to protect humanity from dangerous experimentation but might interfere in necessary and important genetic research with the potential to advance medicine. Supporters of genetic science have therefore attempted to shape public attitudes by promoting successful experiments in the fields of gene therapy and genetic testing, in an effort to highlight the potential of genetic science for medicine and to promote human welfare overall.

Over time, the debate over genetic science has changed as members of the public became more familiar with realistic genetic concepts and ideas, but the older prejudices and fears have remained relevant as well. For instance, in the late 2010s, when it was announced that an overeager Hong Kong scientist had successfully produced babies with edited genomes—a practice illegal in China and many parts of the world—people around the world responded fearfully to the news and many warned that a new era of designer humanity might be at hand, further removing humanity from the natural processes and forces that created the species. Supporters argued that prenatal genetic intervention might prevent diseases and eliminate long-standing genetic defects, but this left the public divided, reigniting mistrust in science and medicine as an industry and escalating fears about how unscrupulous advancement might have unforeseen hazards for humanity.

In many ways, the twenty-first century has been a turning point in genetic science with many genetic techniques once confined to the realm of science fiction becoming scientific reality. This process has kept genetic engineering at the forefront of the scientific debate but is also shaping and changing attitudes about genetics and humanity's experiments with altering the basic building blocks of life. As humanity adjusts to the changing realities of genetic science, however, legislators and public leaders must decide how much legal effort should be invested in protecting humanity from the, still largely hypothetical, dangers of genetic manipulation.

Works Used

Ashworth, James. "Dodo 'De-extinction' Announcement Causes Conservation Debate." *Natural History Museum*. Feb. 15, 2023. www.nhm.ac.uk/discover/news/2023/february/dodo-de-extinction-announcement-causes-conservation-debate.html. Accessed Aug. 2023.

Ball, Philip. *Unnatural: The Heretical Idea of Making People*. New York: Penguin Random House, 2012.

Crichton, Michael. *Jurassic Park*. New York: Arrow Books, 2015.

Luckhurst, Roger. "An Introduction to *The Island of Dr. Moreau*: Science, Sensation, and Degeneration." *British Library*. May 15, 2014. www.bl.uk/romantics-and-victorians/articles/an-introduction-to-the-island-of-doctor-moreau-science-sensation-and-degeneration. Accessed Aug. 2023.

Worthington, Leah. "De-extinction Could Reverse Species Loss: But Should We Do It?" *California*. Cal Alumni Association. Sept. 17, 2021. alumni.berkeley.edu/california-magazine/fall-2021/deextinction-revive-ancient-endangered-species-woolly-mammoths-passenger-pigeons/. Accessed Aug. 2023.

Notes

1. Crichton, *Jurassic Park*.
2. Worthington, "De-extinction Could Reverse Species Loss: But Should We Do It?"
3. Ashworth, "Dodo 'De-extinction' Announcement Causes Conservation Debate."
4. Ball, *Unnatural: The Heretical Idea of Making People*.
5. Luckhurst, "An Introduction to *The Island of Dr. Moreau*: Science, Sensation, and Degeneration."

Should We Bring Back the Dodo? De-extinction Is a Feel-Good Story, but These High-Tech Replacements Aren't Really "Resurrecting" Species

By Risa Aria Schnebly and Ben A. Minteer
The Conversation, March 1, 2023

It's no secret that human activities have put many of this planet's inhabitants in danger. Extinctions are happening at a dramatically faster rate than they have over the past tens of millions of years. An estimated quarter of all species on Earth are at risk of being lost, many within decades.

What can scientists possibly do to stop that trend? For some, the answer is to "de-extinct."

Colossal, a biotechnology company that garnered headlines for its plan to "de-extinct" the woolly mammoth, is now attempting to "bring back" the famously dead dodo bird. The company says its goal is to create a population of undead dodos to put on the Indian Ocean island of Mauritius, where the hefty, flightless creatures lived before humans drove them to extinction in the late 1600s.

As environmental humanists, we study the morality of different conservation interventions, and are interested in how de-extinction might change the ways people think about their responsibilities toward nature. One of us, Ben, is a professor of environmental ethics who explores the ethics of de-extinction in his 2018 book *The Fall of the Wild*. The other, Risa, is a doctoral student researching how de-extinction might change public perceptions about extinction, especially its emotional impact.

What De-extinction Is and Isn't

De-extinction is not exactly what it sounds like. Rather than "bringing back" lost species, it's more of a process to create their high-tech look-alikes.

Scientists would edit the genomes of the dodo's closest living relative—the Nicobar pigeon, which contains the pigeon's full set of DNA—and add some of the most important dodo genes, taken from preserved dodo remains. Then they could put that genome into an egg cell, and let that egg develop into an organism that should look like a dodo.

But that organism wouldn't be genetically identical to the dodo. Nor would it have any other dodos to teach it how to act like and, well, actually be a dodo.

Colossal hasn't successfully created any de-extinct creatures yet. Nor have any other scientists, unless you count the team that cloned the Pyrenean ibex in 2003—but that clone died within minutes. And yet Colossal seems confident, saying it hopes to de-extinct Tasmanian tigers by 2025 and woolly mammoths by 2027. They're certainly amassing a fortune to make it happen: Since its founding in 2021, Colossal has raised over US$225 million from tech investors, Paris Hilton and even a CIA-backed venture capital firm.

Possibilities, or Pitfalls?

Supporters have argued that de-extinction will eventually help restore ecosystems. "Bringing back" passenger pigeons could help restore forests in the northeastern United States, for example, while woolly mammoth proxies could help restore the Siberian steppe and keep permafrost frozen. Some de-extinction advocates have also positioned their projects as potential long-term solutions to combating mass biodiversity loss in general.

But many ecologists and ethicists have highlighted the uncertainty around introducing these novel creatures into the wild. Even if the de-extinct dodos did act more or less like their extinct counterparts, it's hard to know how a habitat that hasn't had any dodolike birds in it for 350 years would be affected by this new species. Opponents have pushed back even more strongly against claims that de-extinction could be a widespread solution, pointing out how bringing back one species at a time would not be enough to curb the Earth's losses.

Other issues include how to decide where all these de-extinct creatures would live, as well as animal welfare concerns: for potential surrogate animals that would be impregnated, and the de-extinct creatures themselves, which never asked to be "brought back."

More Than Science

To us, one of the more interesting questions about de-extinction has to do with how it changes the way people think about extinction.

Some de-extinction boosters have argued that de-extinction could create a more hopeful story about humans' ability to combat mass extinction. Many others share the desire for more inspiring conservation stories, too. Some conservationists and psychologists have argued that environmentalists need more positivity to get people engaged with environmental issues.

Others, however, say de-extinction isn't hopeful, but misleading. Many worry that de-extinction actually risks making humans less inclined to care about ongoing extinctions. After all, why care about preventing extinction if we can eventually reverse it?

It's hard to rally the troops with a message of unrelenting guilt and despair. But reckoning with those difficult emotions can be useful for reflecting on humanity's responsibilities—especially considering that extinction is our fault to begin with, and since de-extinction isn't really "resurrecting" anything.

> **Many worry that de-extinction actually risks making humans less inclined to care about ongoing extinctions. After all, why care about preventing extinction if we can eventually reverse it?**

In fact, some scholars argue that what humans really need is to learn to grieve extinct species. Grief, they say, is a transformational process that helps people recognize the value of what's been lost and appreciate what's left. Grief will never be enough without action. But we believe learning how to grieve together can be a more responsible and honest way to cope with extinction than pretending it can simply be undone.

So which is better at motivating care for the environment: positive or negative stories? There are still no sure answers, and testing their impact on audiences today is a key part of Risa's research. Perhaps it can help conservationists at large learn how to tell more motivational stories—but it will take some time to get there.

In the meantime, we suggest that de-extinction scientists and advocates call de-extinction what it really is: not resurrecting extinct species, but creating their replacements.

Print Citations

CMS: Schnebly, Risa Aria, and Ben A. Minteer. "Should We Bring Back the Dodo? De-extinction Is a Feel-Good Story, but These High-Tech Replacements Aren't Really 'Resurrecting' Species." In *The Reference Shelf: Gene Editing & Genetic Engineering*, edited by Micah L. Issitt, 115–117. Amenia, NY: Grey House Publishing, 2023.

MLA: Schnebly, Risa Aria, and Ben A. Minteer. "Should We Bring Back the Dodo? De-extinction Is a Feel-Good Story, but These High-Tech Replacements Aren't Really 'Resurrecting' Species." *The Reference Shelf: Gene Editing & Genetic Engineering*, edited by Micah L. Issitt, Grey House Publishing, 2023, pp. 115–117.

APA: Schnebly, R. A., & Minteer, B. A. (2023). Should we bring back the dodo? De-extinction is a feel-good story, but these high-tech replacements aren't really "resurrecting" species. In Micah L. Issitt (Ed.), *The reference shelf: Gene editing & genetic engineering* (pp. 115–117). Amenia, NY: Grey House Publishing. (Original work published 2023)

Why Genetic Engineering Experts Are Putting a Spotlight on Victoria Gray's Case

By Rob Stein
NPR, March 7, 2023

Victoria Gray's life has been transformed by her treatment for sickle cell disease with the gene-editing technique called CRISPR. She's in London telling her story at a scientific summit.

A MARTÍNEZ, HOST: Some of the world's most celebrated experts on genetic engineering are in London this week to debate the promise and the peril of gene editing. Yesterday, the summit put the spotlight on one person, Victoria Gray. The Mississippi woman was the first person with sickle cell disease to be treated with a gene-editing technique known as CRISPR. NPR health correspondent Rob Stein had exclusive access to chronicle Gray's experience. Here he is with her in London.

ROB STEIN, BYLINE: When I first met Victoria almost four years ago, she was lying in a hospital bed in Nashville, so weak she could barely get out of bed. She'd been tormented by the devastating blood disorder her whole life and had just gone through a grueling procedure to have billions of her own bone marrow cells genetically modified and infused back into her body. When we meet at her hotel in London on Sunday, Victoria looks like a different person.

Hi, Victoria.

VICTORIA GRAY: Hi, Rob.

STEIN: How are you?

GRAY: I'm doing good. How about you?

STEIN: I'm great. That's great. It's so good to see you.

GRAY: It's good to see you, too.

STEIN: She had just arrived in London with her husband, Earl, from her hometown in Mississippi—her first trip outside the U.S. ever. And even though she didn't sleep much on the overnight flight, she can't wait to see the sights.

Are you ready for this, Victoria?

GRAY: Yes, I am.

STEIN: You sure?

GRAY: I'm excited.

STEIN: OK.

GRAY: Yes, I'm sure.

STEIN: Here we go.

Before the treatment, deformed red blood cells would incapacitate her with horrible, unpredictable attacks of pain, sending her rushing to the hospital for pain medication and blood transfusions. She could barely get out of bed many days, struggled to care for her four children, keep a job. Today, at 37, all of her symptoms have disappeared. She works full-time as a Walmart cashier, keeps up with her teenagers, so she thought she could handle exploring the city. We head out to find the British Museum.

She was the first person with sickle cell disease to be treated with a gene-editing technique known as CRISPR. Before the treatment, deformed red blood cells would incapacitate her with horrible, unpredictable attacks of pain, sending her rushing to the hospital for pain medication and blood transfusions. Today at 37, all of her symptoms have disappeared.

GRAY: I would have never been able to walk this long before.

STEIN: How much of a difference is it?

GRAY: It's a huge difference—like night and day.

STEIN: I can't imagine having lived a whole life in one way and then having suddenly be so much better.

GRAY: Yes, especially when you have a disease that they say was incurable. So I'm here now. I feel like I got a second chance.

STEIN: We finally make it to the museum. Victoria is not thrilled by the mummies, but I find her studying a small wooden artifact hanging on the wall.

GRAY: It's nice seeing all the old artifacts, especially the cross.

STEIN: Why especially the crosses?

GRAY: Because religion is something that I hold close to my heart. My faith is what brought me this far. And God did his part—you know?—for what I prayed about for years. And together, hand in hand, God and science worked for me.

STEIN: Next stop is the London Eye—a huge Ferris wheel that towers over the city. Victoria's keen for a ride, even though she's afraid of heights. We've been exploring for hours. We climb on board and circle to the top.

How are you feeling, Victoria?

GRAY: I still feel good.

STEIN: What do you think of this view?

GRAY: Oh, it's a beautiful view.

STEIN: So that looks like Big Ben right there. You see it?

GRAY: Yes.

STEIN: Did you ever think you'd be able to get a view like this?

GRAY: No, I didn't. Part of my dream's coming true.

STEIN: As we circle back down, I ask Victoria how she's feeling about addressing the international gene editing summit the next day.

GRAY: So I'm very excited—a little nervous, honestly, speaking in front of a large crowd.

STEIN: So what do you think will be more nerve-wracking—doing the London Eye or speaking at that summit?

GRAY: Speaking at the summit, of course.

UNIDENTIFIED ANNOUNCER: OK. Our flight is almost over, so please make sure you take everything with you when you leave.

STEIN: The next morning, Victoria makes her way through the crowd at the summit and finds a seat in the auditorium as Robin Lovell-Badge opens the three-day meeting at the Francis Crick Institute.

ROBIN LOVELL-BADGE: Hello, everyone. I'm very pleased to see so many people here. So welcome to the third Summit on Human Genome Editing.

STEIN: Speaker after speaker described the latest scientific advances in gene editing. David Liu from Harvard sounded a bit echoey because he addressed the summit remotely.

DAVID LIU: There are more than 200 patients to date, including Victoria, Patrick and Carlene, pictured here, that have been treated in clinical trials with CRISPR nucleases targeting DNA sequences that, when disrupted, offer clinical benefit. You'll hear more from Victoria about her experience directly later today.

STEIN: Finally, it's Victoria's turn on the podium.

GRAY: Good evening. I'm Victoria Gray, and I'm a 37-year-old mother of four and a sickle cell survivor. Take a moment to go on a journey with me.

STEIN: For 10 minutes, Victoria repeatedly chokes back tears as she describes her life with sickle cell, including one especially torturous pain crisis...

GRAY: During this hospital stay, with a ketamine infusion in one arm and a Dilaudid infusion in the next, but still no pain relief, I called all the doctors into the room, and I told them that I could no longer live like this. I went home, and I continued to pray and look to God for answers.

STEIN: ...And how she finally received the CRISPR gene-edited cells—super cells, she calls them—as part of a study.

GRAY: The life that I once felt like I was only existing in I am now thriving in. I stand here before you today as proof that miracles still happen and that God and science can coexist. Thank you for allowing me to share my story with you.

STEIN: As Victoria walks off the stage, the audience of scientists, doctors, bioethicists and others gives her a standing ovation. Vertex Pharmaceuticals and CRISPR Therapeutics, which sponsored the study Victoria volunteered for, are asking the Food and Drug Administration to approve the treatment. That would make it the first gene-editing therapy to become widely available.

But for the rest of the afternoon, speakers warned that there are still big questions about all this. How many patients will it help? How long will it last? The treatment is complicated and is expected to be really expensive—possibly costing millions of dollars. Will it be available to the patients who need it most, especially in less affluent countries, where sickle cell is most common?

Print Citations

CMS: Stein, Rob. "Why Genetic Engineering Experts Are Putting a Spotlight on Victoria Gray's Case." In *The Reference Shelf: Gene Editing & Genetic Engineering,* edited by Micah L. Issitt, 118–121. Amenia, NY: Grey House Publishing, 2023.

MLA: Stein, Rob. "Why Genetic Engineering Experts Are Putting a Spotlight on Victoria Gray's Case." *The Reference Shelf: Gene Editing & Genetic Engineering,* edited by Micah L. Issitt, Grey House Publishing, 2023, pp. 118–121.

APA: Stein, R. (2023). Why genetic engineering experts are putting a spotlight on Victoria Gray's case. In Micah L. Issitt (Ed.), *The reference shelf: Gene editing & genetic engineering* (pp. 118–121). Amenia, NY: Grey House Publishing. (Original work published 2023)

Seven Times Science Fiction Got Genetic Engineering Right

By Sue Burke
TOR, June 1, 2021

We love to tinker with our environment, especially with other life forms. We try to change them to suit our needs, using every tool we can find or invent. Science fiction goes one step further, imagining tools we haven't invented yet and doing things that don't seem possible. Yet sometimes science fiction's impossible dreams have echoed real-life tinkering—even when our imaginations birthed nightmares...

We could say genetic engineering started in 1926 when Thomas Hunt Morgan discovered the role chromosomes play in heredity. Or in 1953 when James Watson and Francis Crick (with Rosalind Franklin) described the double-helix structure of DNA.

I contend that we actually started genetic engineering thousands of years ago using selective breeding. Since Mesolithic times, we've successfully changed plants and animals in profound ways. We didn't know why it worked, but we knew we had the power to transform life, and we never stopped using that power in real life or in our imagination.

Here are seven ways sci-fi writers correctly predicted what genetic engineering could do.

We Will Make Monstrous Changes in Animals

H.G. Wells wrote *The Island of Doctor Moreau* in 1896, describing ghastly combinations of animals with other animals, and of animals with humans. He was inspired in part by the horrors of vivisection, an important social issue of his time. In the novel, Doctor Moreau creates chimeras, or cross-species combinations, including bear-dog-oxen, hyena-swine, mare-rhinoceros, ape-man, leopard-man, swine-man, swine-woman, wolf-man, wolf-woman, and dog-man through brutal surgeries. Eventually it all leads to disaster.

In our own time, using the full powers of genetic engineering, we're combining animals, such as mouse-rat, sheep-goat, chicken-quail, and human-pig. Most recently, Tao Tan, a biologist at Kunming University of Science and Technology, with the help of a large team, made part-monkey, part-human embryos. What could possibly go wrong? We'll find out.

(Just to be clear, a turducken is not the result of genetic engineering. It involves culinary engineering.)

We Will Make Monstrous Changes: The Sequel

In 1990, Michael Crichton brought dinosaurs back to life in *Jurassic Park,* and the plot hinges on a fictional misjudgment in the genetic engineering. Gaps in dinosaur genes are spliced with reptilian, avian, or amphibian DNA. To control the dinosaur population, only females are bred, but it turns out that frogs can sometimes change from female to male. Oops. Those and other errors mean the dinosaurs eventually escape.

Crichton was inspired by genetic engineering, still new in 1990, but we've done amazing things in the past with selective breeding. About 9000 years ago, people in what is now southern Mexico began to experiment with a kind of grass called teosinte.

It protects its seeds with a hard casing. Ancient agriculturalists slowly rebuilt it into maize (corn). The seed casings became the central cob, and the luscious seeds were exposed to predators like us.

Another example: around 23,000 years ago, we started changing wolves into dogs. Now we've gone so far as to make miniature chihuahuas. These may not be actual monsters, but tiny chihuahuas and corn on the cob illustrate what horrors we could create if we tried. Even simple genetic tools hold great power, which comes with great responsibility.

Genetic Engineering Will be Dehumanizing

This is a common theme in science fiction. *The Windup Girl* by Paolo Bacigalupi is a good example. The "windup girl" is not a human. She's one of the New People, engineered and creche-grown, considered soulless beings, perhaps devils. They toil as slaves, soldiers, and toys.

We can easily accept the novel's premise because in real life, we've tried many times to define some people as more human than others on the basis of such differences as skin color, gender, religion, or national origin. Every time, disaster followed.

In general, we haven't tried genetic engineering on human beings, but one instance of reverse engineering stands out. The dangers of inbreeding have long been understood, but greed can overcome good sense. During Renaissance times, the House of Habsburg in Europe intermarried to hold onto power, eventually resulting in King Charles II of Spain (1661-1700). He was so inbred he could barely eat, speak, or walk. That mattered little. He was a thing to occupy a throne, providing other people with agency. The institution of royalty itself might be dehumanizing.

Accidents Won't Always Be Bad

In Adrian Tchaikovsky's 2015 novel *Children of Time*, various creatures are accidentally genetically uplifted, in particular spiders. The spiders slowly evolve in intelligence and become heros, willing to fight to protect the weak and to risk their lives to save others—big, arachnophobia-inspiring heroes. In the meantime, humans engage in continued, senseless self-destruction. We don't seem to be the smartest species in the story.

In real life, we also stumble into lucky accidents. People in Mesopotamia domesticated sheep at least 10,000 years ago for meat, but the change to the gene that made the animals more docile also had an unanticipated side effect. It made the fleece start to crimp.

In our own lives, we have a current example of genetic engineering doing good: the Pfizer and Moderna vaccines against Covid-19 are saving lives.

Soon, it could be spun into wool.

As a result, 6,000 years ago, Babylonians were wearing woven woolen clothing as a proud sign of civilization.

Genetic Engineering Will Have Great Potential for Evil

In the 2017 novel *Borne* by Jeff VanderMeer, a city is destroyed by genetically engineered monsters, half-creatures, and ambiguous beasts. Giant flying bears, strange anemone-like blobs, compost worms, memory beetles, and other creatures populate this horrible future. The disaster—a Collapse worthy of a capital C—was birthed by unhinged corporate avarice.

In our own consensus reality, corporate involvement in genetic engineering has generated all kinds of controversy, but I want to point to one instance in which corporate avarice is beyond debate. Tobacco companies have genetically engineered tobacco to be more addictive. Mic drop.

Genetic Engineering Will Have Great Potential for Good

Success can be harder to write than dystopia, so *Lilith's Brood* by Octavia E. Butler, published in 2000, needed three novels to reach a happy ending. Eventually, humans and an alien species called the Oankali find ways to live together—really together. Along the way, the trilogy explores complex themes related to genetic engineering, such as identity, social integration, power, and eugenics.

In our own lives, we have a current example of genetic engineering doing good: the Pfizer and Moderna vaccines against Covid-19 are saving lives. They use a specific kind of mRNA that makes a few of our cells reproduce the Covid spike protein, a specific fragment of the Covid virus. When our immune system sees those spikes, it builds antibodies and T-cells to fight them. The vaccine doesn't re-engineer our DNA, but the science behind genetic engineering provided the knowledge base for the very rapid development of the vaccines. (I'm team Pfizer.)

Genetic Engineering Will Pose a Clear and Present Danger

Many science fiction stories portray disaster, including the 2003 novel *Oryx and Crake* by Margaret Atwood. It shows how uncontrolled genetic engineering can destroy humanity—intentionally.

Right now, our technical ability to deliberately create a harmful organism, micro-organism, or virus seems limited, but sooner or later we will have that power. What are we doing about it? So far, 183 countries have signed the Biological Weapons Convention, which bans the use of disease-causing organisms or toxins to harm or kill humans, animals, or plants. Signatory countries are required to control the actions of corporations and research organizations under their jurisdiction. Good luck with that. Meanwhile, non-state actors, such as terrorist groups, have little incentive to sign this sort of agreement.

We can barely control other kinds of weapons of mass destruction. Fully 191 countries have signed the Treaty on the Non-Proliferation of Nuclear Weapons. Four of the countries that did not sign either have nuclear weapons or want them, and one signatory country is currently in non-compliance. Worse than that, about 3,750 nuclear warheads are active right now, and 1,800 remain in a state of high alert.

Given our minimal success with controlling nuclear weapons, we might want to think harder about biological weapons. Even very simple biological engineering techniques have reshaped our world. We now possess advanced engineering, and only its technical difficulty has kept us safe so far. It will get easier to use. Science fiction has long been warning us that time is running out, and even its wildest ideas keep coming true.

Print Citations

CMS: Burke, Sue. "Seven Times Science Fiction Got Genetic Engineering Right." In *The Reference Shelf: Gene Editing & Genetic Engineering*, edited by Micah L. Issitt, 122–125. Amenia, NY: Grey House Publishing, 2023.

MLA: Burke, Sue. "Seven Times Science Fiction Got Genetic Engineering Right." *The Reference Shelf: Gene Editing & Genetic Engineering*, edited by Micah L. Issitt, Grey House Publishing, 2023, pp. 122–125.

APA: Burke, S. (2023). Seven times science fiction got genetic engineering right. Micah L. Issitt (Ed.), *The reference shelf: Gene editing & genetic engineering* (pp. 122–125). Amenia, NY: Grey House Publishing. (Original work published 2021)

How Real Is Genetic Engineering in Sci-Fi?

By Courtney Linder
Popular Mechanics, August 26, 2020

In the 2002 film *Spider-Man*, a conspicuous, genetically modified spider bites Peter Parker on the hand, giving him qualities like 20/20 vision and the ability to shoot webs from his wrists. Then there's the 1993 blockbuster *Jurassic Park*, in which Dr. Grant and Dr. Satler use dinosaur DNA, partially preserved in amber, to bring the likes of velociraptors and brachiosauruses back to life.

While divergent in approach and interpretation, both movies use genetic engineering as the underpinning for serious conflict. Since the beginning of modern genetic engineering in the 1970s, sci-fi has grappled with some of the most probing questions about the technology: Is genetic engineering ethical, can it fundamentally improve human life, and most significantly, what are some of the logical worst-case scenarios?

Let's use several classic examples of genetic engineering in sci-fi to separate fact from fiction.

Genetic Engineering 101

Herbert Boyer and Stanley Cohen created the first successful genetically engineered organism in 1973. They devised a biotechnology method that involved cutting out genes from one organism and inserting them into another. In this case, they took a gene that encodes antibiotic resistance and combined it with a separate strain of bacteria, giving the latter antibiotic properties.

Rudolf Jaenisch and Beatrice Mintz formalized the process for animals the following year. In the earliest experiments, that meant combining outside DNA with mouse embryos, according to Science In the News, a group of graduate science communicators at Harvard University.

By 1980, major firms began to rely on genetic engineering. In particular, General Electric created a new kind of bacteria, designed to break down crude oil for use in oil spill cleanups. Because the U.S. Supreme Court allowed GE to patent the bacteria, it created a precedent for companies that wanted to legally own genetically modified organisms, or GMOs. These are plants, animals, or other forms of life that scientists have manipulated to attain certain traits.

Today, one of the most profitable and common use cases for GMOs is in agriculture. The very first experiments using genetically modified food crops dates back to 1987, when Calgene Inc., a Davis, California-based biotech company, introduced its "Flavr Savr tomato." The firm modified the tomatoes to be more firm and to last longer. In a first, the U.S. Department of Agriculture approved the GMO food product for production.

Soybeans are now one of the most common examples of GMO crops in the U.S. According to the U.S. Food and Drug Administration, GMO soybeans made up 94 percent of all soybeans

> **Movies use genetic engineering as the underpinning for serious conflict.**

planted in 2018. Like most other genetically modified plants, the soy is used as feed for animals like poultry and livestock, as well as an additive in processed foods.

Beyond that, scientists use genetic engineering to make animals grow larger and develop new pharmaceuticals. CRISPR ("Clusters of Regularly Interspaced Short Palindromic Repeats") is a relatively new tool in genetic engineering that allows scientists to more easily select and alter genes by snipping out strands of DNA.

In some circles, GMOs get a bad rap, with entire conspiracy theories dedicated to the technology. Perhaps some of this can be explained through the lens of science fiction, which examines some of the most damning possibilities if genetic engineering were to go horribly wrong. However, these folks would do well to remember the "fiction" component of the genre.

Fact or Fiction?

Not all sci-fi is created equally—some films and TV shows have more realistic depictions of genetic engineering than others. Here are some of our favorite examples in popular culture, from the most accurate adaptions to some frankly out-there interpretations.

Blade Runner (1982)

What's proposed?

In the year 2019, Los Angeles is thick with pollution and overpopulation. The Replicants—a group of genetically engineered humans, meant to be perfect—are more attractive, smarter, and stronger than their counterparts. However, because they were incubated in the lab, the replicants have a short lifespan of just four years. A special police force hunts down these quasi-humans, identifying them from regular humans with a series of questions of physical tests, known as the Voight-Kampff Test.

How realistic is it?

It's a bit out there. "Although we have been able to identify and manipulate individual genes, we still have a limited understanding of how an entire human emerges from genetic code," Fumiya Iida, a lecturer in mechatronics at the University of Cambridge, wrote in *The Independent* in 2017. "As such, we don't know the degree to which we can actually [program] code to design everything we wish."

But there is some degree of realism in the genetic engineering: the kill-switch that causes the Replicants to die after four years is not completely unlike some lab-modified cells. In T-cell therapy, for example, researchers have programmed in the ability to turn the cells on or off if they go rogue.

Star Wars: Episode II—Attack of the Clones (2002)

What's proposed?

On the ocean planet of Kamino, Jedi Knight Obi-Wan Kenobi finds out the Republic has been working on an army of clones, based on the genetic makeup of bounty hunter Jango Fett.

How realistic is it?

Twins are perfect clones of one another, since they have identical genes. But the idea of creating a perfect "twin" in the lab is extremely controversial, and thus, regulatory bodies have blocked this kind of work. The International Society for Stem Cell Research opposes "reproductive" cloning, according to *USA Today.*

This is in large part because cloning would almost certainly lead to embryos that could not live through a pregnancy, or would at least be born sick or with disabilities. So the *Star Wars* theory probably wouldn't pan out for a top-tier army.

The Fly (1986)

What's proposed?

In his lab, scientist Seth Brundle has a set of teleportation pods that can move inanimate objects, but drastically mutilate live tissue in animals. After some tinkering, he discovers the machines only create synthetic versions of the object in question, rather than truly recreate the object itself. He tries teleporting himself, only to enter the machine with a housefly. Over time, Brundle starts to see increased sexual stamina and strength, eventually realizing he's fused himself with a fly at the genetic level.

How realistic is it?

More than you think! It turns out humans share about 70 percent of their DNA with the common housefly. "The genes that we share with them are responsible for embryonic development," DNA expert Erica Zahnle told the *Chicago Tribune* in 2017.

Still, that doesn't mean scientists can necessarily combine humans and flies. It's problematic at the cellular level because humans have 23 pairs of chromosomes, and flies only have six. What's happening in *The Fly* is more like a viral infection than a viable example of genetic engineering: a fly virus invades the human cell and makes copies of itself. Because there's no splicing or sharing of genes, it wouldn't lead to the gruesome creature Jeff Goldblum becomes.

Jurassic Park (1993)

What's proposed?

Scientists working on a Costa Rican island have come up with a way to bring dinosaurs back to life in a bid to create a theme park. After locating dinosaur DNA inside mosquitoes that were preserved in amber, the researchers combined the partly damaged genetic material with frog DNA. All of the animals were made female to avoid non-test tube breeding.

How realistic is it?

Some of the film's science talk is based in reality, but the applications are, at best, like stretching the truth. For one thing, amber *is* great at preserving organic materials, but not for 65 million years back to the age of dinosaurs. DNA sequences long enough to be usable and readable for cloning purposes probably can't survive more than 1.5 million years, according to recent research.

"We can find fragments of DNA [in dinosaur fossils], but not enough of a genome to activate it," Mark Norell, Chair and Macaulay Curator in the American Museum of Natural History's Division of Paleontology, told *Popular Mechanics* in 2018. "The chances of finding a whole genome is nearly impossible."

Spider-Man (2002)

What's proposed?

During a school field trip to Columbia University's science department, a blue and red spider bites Peter Parker on the hand. It turns out this specimen is one of 15 genetically modified arachnids that scientists created from the synthesized RNA from three natural species of spider. Long story short: Parker ends up with super powers, like the ability to build webs and swing from skyscrapers.

How realistic is it?

Not very, according to a May 2002 article in *Genome News Network*. "The mechanism in the movie is pure fantasy," Jonathan Coddington, then a research scientist at the National Museum of Natural History, told the news source. Because spider venom contains mostly globular proteins (a type of spherical proteins including hemoglobin and insulin), the transfer of genetic material from spider to human could only occur if a viral parasite were carrying all of the spider's most incredible genes.

Annihilation (2018)

This content is imported from YouTube. You may be able to find the same content in another format, or you may be able to find more information, at their web site.

What's proposed?

After a meteor lands inside a lighthouse along the southern coast of the U.S., the government creates a restricted area around it, known as Area X. Anyone who enters this dangerous zone called "The Shimmer" dies. Mutated plants and animals abound, including an albino alligator with teeth like a shark, and a bear that screams with the human voice of its last meal. It turns out The Shimmer is refracting plant and animal DNA, leading to mutations.

How realistic is it?

The lines between species are, as the title suggests, annihilated. This isn't completely unrealistic, considering scientists have introduced advanced chimeras in the last few years, including a mouse with 4 percent human DNA. Meanwhile, the forces in The Shimmer that cause organic DNA to scramble are actually found in nature. Known as horizontal gene transfer, this process is the uptake of DNA from one organism into the cells of another.

"The first organism's DNA is incorporated into the new organisms' chromosome and then actually functions, changing the properties of the receiving organism," Jeffrey Way and Pamela Silver—two scientists at Harvard's Wyss Institute for Biologically Inspired Engineering—wrote in *Sloan Science & Film* after the movie debuted. Still, some of the most prevalent examples of horizontal gene transfer include the formation of mitochondria and chloroplasts.

In other words, this is totally rare and won't lead to a horrifying human-bear anytime soon.

Print Citations

CMS: Linder, Courtney. "How Real Is Genetic Engineering in Sci-Fi?" In *The Reference Shelf: Gene Editing & Genetic Engineering*, edited by Micah L. Issitt, 126–130. Amenia, NY: Grey House Publishing, 2023.

MLA: Linder, Courtney. "How Real Is Genetic Engineering in Sci-Fi?" *The Reference Shelf: Gene Editing & Genetic Engineering*, edited by Micah L. Issitt, Grey House Publishing, 2023, pp. 126–130.

APA: Linder, C. (2023). How real is genetic engineering in sci-fi? In Micah L. Issitt (Ed.), *The reference shelf: Gene editing & genetic engineering* (pp. 126–130). Amenia, NY: Grey House Publishing. (Original work published 2020)

A History of Genome Engineering in Popular Culture

By Kartik Lakshmi Rallapalli
AddGene Blog, February, 2020

The revolution in genetic engineering techniques is a speculation of yesteryear which has been realized recently. Science Fiction (SciFi) writers have been curious about the capability of transforming the genetic code of living organisms and its societal implications even before the discovery of genes themselves.

How has scientific progress in genetic editing impacted the world of fiction and vice versa? We draw a parallel between the timelines of the scientific advancements and the noteworthy work of fiction that it inspired. CAUTION: Spoilers alert!

Pre-DNA Era

H.G. Wells, the father of scientific fiction, was also a trained biologist. This is apparent in his work *The Island of Dr. Moreau* (1896) where he explores the concepts of animal testing and modifications. The premise of the novel involves Dr. Moreau who creates human-animal hybrids. This book is published 30 years after the seminal work of Gregor Mendel on the genetic inheritance (1866) and just when the Mendelian genetics was being revived by the botanist Hugo de Vries (1896).

In 1928, Frederick Griffith used mouse models to prove that *Streptococcus pneumoniae* bacteria could be transformed from one organism to the other. This is now famed as the Griffith's experiment.

DNA Era

The period between 1940-1960 was pivotal for genetics and SciFi. Scientists around the world began looking for the magic molecule that serves as the code of life. In 1944, Oswald Avery, Colin MacLeod, and Maclyn McCarty's experiment debunked the myth that protein is the carrier of genetic information and led to the coronation of DNA as the molecule of life. This was further reinforced by the findings of the 1952 Hershey-Chase experiment. In the same year, Rosalind Franklin and her student Raymond Gosling took the first ever X-ray diffracted image of the DNA. Finally, in 1953, Franklin along with James Watson and Francis Crick resolved the DNA double helix. In the 1960s, Har Gobind Khorana, Marshall W. Nirenberg, and Robert W. Holley cracked the genetic code and explained how the four letters of the DNA get translated into the twenty amino acids. Once scientists established that DNA encoded all the information necessary for regulating and propagating life, there was no turning back.

While the world of molecular genetics was booming, SciFi saw the rise of literary giants such as Isaac Asimov, Arthur C. Clarke, James Blish, and John W. Campbell. This period is termed as the golden age of SciFi, but the common themes that these authors were exploring were space and robots. Warping to stars in space ships and the rise of killer machines that exterminate entire humanity were the two most popular topics that sparked the imagination of most of the authors during this period.

Genetic Engineering Era: Pre-CRISPR

Having visualized and understood the DNA, the next milestone for the molecular biologist was to devise methods to manipulate it. The term genetic engineering itself is attributed to SciFi writer Jack Williamson who used it in his 1951 novel *Dragon's Island*. In 1976, the world's first genetic engineering company, Genentech, was founded.

This commercialization of gene editing technology led to the development of many important medical and agricultural products. But it also made SciFi director Ridley Scott wonder what would happen if these corporations would extrapolate genetic engineering to exploit human beings. Ridley Scott directed *Blade Runner* (1982). It is a dystopian film that revolves around synthetic humans called replicants which are genetically engineered by a powerful corporation to serve as slaves.

Meganucleases

In 1989 the first ever DNA-editing enzymes, meganucleases, were discovered. These proteins are characterized by large DNA recognition sites, which makes them highly specific.

Jurassic Park by Michael Crichton was published in 1990, just a year after the discovery of meganucleases. The storyline involves the resurrection of dinosaurs by extracting the DNA from fossilized amber and supplementing it with the DNA of present-day frogs via genetic engineering. The corresponding movie came out in 1993 and sequels of *Jurassic Park* (1997, 2001, 2015, and 2018) were highly successful in the box office, indicating the insatiable appetite of the moviegoers for dinosaurs wreaking havoc in a theme park.

Released in 1997, *GATTACA* is an American SciFi film starring Ethan Hawke, Uma Thurman and Jude Law. The movie's name is itself a combination of the four bases of DNA—A, T, G, and C. In this movie Ethan Hawke's character revolts against class divide that is created by genetics.

Zinc Finger Nucleases

Zinc finger nucleases (ZFNs) are artificial restriction enzymes generated by fusing a zinc finger DNA-binding domain to a DNA-cleavage domain (FokI). Zinc finger nucleases were first designed in 2005. The DNA-binding domain can be engineered to target specific DNA sequences while FokI then induces a double strand break. These breaks can then be repaired by homologous recombination with a repair template. Each zinc finger recognizes a nucleotide triplet and when several ZFNs are strung together, it can read many base pairs simultaneously to increase specificity. This modular approach made ZFNs highly appealing to researchers.

> **Once scientists established that DNA encoded all the information necessary for regulating and propagating life, there was no turning back.**

TALENs

In 2010, transcription activator-like effector nucleases (TALENs) were engineered. TALENs follow a DNA-binding and DNA-cleaving strategy similar to the ZFNs, but instead of recognizing nucleotide triplets, the components recognize individual nucleotides. This imparts TALENs more specificity and a broader substrate preference than the ZFNs.

Genetic Engineering Era: Post CRISPR

The strategy of using a protein to recognize DNA sequences is not ideal. For the early generation gene editors, new amino acid recognition sites had to be designed each time we wanted to target new genomic sites. The discovery of the Cas proteins in 2005 was a major breakthrough in the world of genetic engineering as CRISPR-Cas editing soon replaced the protein-based DNA-binding domain with a single-guide RNA sequence. The Cas proteins, just like its predecessor induce a double-strand break at the target genomic location. This is typically followed by non-homologous end-joining that results in gene knockouts.

Star Trek: Into Darkness (2013) which is the remake of the space-melodrama *Star Trek II: The Wrath of Khan* (1982) combined the worlds of genetic engineering and space travels. The movie sees the USS *Enterprise* crew fight against a genetically engineered superhuman Khan.

The novel, *Inferno* (2013) and the movie version (2016) stars Tom Hanks as the protagonist (Langdon). The plot involves Langdon chasing an anti-villain geneticist named Zobrist who designs a virus that can potentially kill one-third of the human beings (just like the black plague).

Base Editors

The major drawback of all the endonuclease-based gene editors (meganucleases, ZFNs, TALENs, and CRISPR-Cas) is that they induce double-strand breaks (DSBs) in the genome. These technologies rely on the cellular DNA repair machinery, which can lead to high frequency of indel formation in the genome and are inefficient at introducing point mutations. In 2016, the researchers at the Broad Institute envisioned a new strategy to introduce point mutations in the genome.

They fused a catalytically impaired Cas9 protein to a DNA-editing enzyme to generate the prototypical base editor enzymes. The cytidine base editors (CBE) use a cytidine deaminase to introduce a C→T change at precise locations in the genome. This was soon followed by the adenine base editor (ABE) enzyme in 2017. ABEs rely on a similar strategy as CBE but use a ssDNA adenosine deaminase (evolved from a tRNA deaminase enzyme, TadA) to induce a A→G mutation at the target genomic loci. Together, CBE and ABE have obviated the indel formation rates that are typically associated with the DSBs and offered unprecedented prospects of curing diseases caused due to single nucleotide variations in the genome.

Due to the rise of CRISPR in genetic engineering and open advertisement of the technology on various non-scientific platforms, CRISPR has also transcended into the world of SciFi. This is apparent in the use of the word 'CRISPR' for anything associated with genetic engineering in the recent SciFi movies. *Rampage* (2018), starring Dwayne "The Rock" Johnson, calls out CRISPR as the genetic engineering experiment which goes horribly wrong. The plot of the movie involves The Rock and his super giant albino gorilla fighting off a monstrous wolf and crocodile, that were "affected by CRISPR."

Prime Editing

The latest breakthrough in genome editing is prime editing, which was designed in 2019. It's based on a prime editor which is a fusion between Cas9 and reverse transcriptase. Using an elongated template + guide RNA called pegRNA (prime editing RNA), the prime editor can install all possible point mutations, delete short stretches of DNA and even insert new DNA at the target genomic loci.

A Last Word on the Depiction of Genome Engineering in Pop Culture

SciFi is one of the most imaginative and most important genres of literary fiction. It has always served as a vision of the future and to make us reframe our perspective of the current. Although genome editing technologies have brought a renaissance in the field of health sciences, energy, and agriculture, its portrayal in popular fiction has been largely negative. But these negative depictions serve as cautionary tales for the entire scientific community and remind us to always consider the ethics and broader implications of our research.

Print Citations

CMS: Rallapalli, Kartik Lakshmi. "A History of Genome Engineering in Popular Culture." In *The Reference Shelf: Gene Editing & Genetic Engineering*, edited by Micah L. Issitt, 131–135. Amenia, NY: Grey House Publishing, 2023.

MLA: Rallapalli, Kartik Lakshmi. "A History of Genome Engineering in Popular Culture." *The Reference Shelf: Gene Editing & Genetic Engineering*, edited by Micah L. Issitt, Grey House Publishing, 2023, pp. 131–135.

APA: Rallapalli, K. L. (2023). A history of genome engineering in popular culture. In Micah L. Issitt (Ed.), *The reference shelf: Gene editing & genetic engineering* (pp. 131–135). Amenia, NY: Grey House Publishing. (Original work published 2020)

Americans Are Closely Divided Over Editing a Baby's Genes to Reduce Serious Health Risk

By Lee Rainie, Cary Funk, Monica Anderson, and Alec Tyson
Pew Research Center, March 17, 2022

Americans strongly support using gene editing techniques for people's therapeutic needs. But, when it comes to their potential use to enhance human health over the course of a lifetime by reducing a baby's risk of getting serious diseases or conditions, as many Americans think this would be a bad idea for society as say it would be a good idea. The public is also closely divided over whether they would want this for their own baby. As with previous Pew Research Center surveys on this topic, women and more religious Americans are less accepting of gene editing for this purpose.

Scientific advances in the use of CRISPR technology are expanding the possibilities for using gene editing. These techniques are currently in development for therapeutic needs. Clinical trial data suggests that gene therapy can be effective in treating some heritable blood disorders such as sickle cell anemia. Other trials have shown promise for treatment of life-threatening rare diseases.

Public divided over societal impact of using gene editing for babies to reduce risk of disease

% of U.S. adults who say the widespread use of gene editing to greatly reduce a baby's risk of developing serious diseases or conditions would be a ___ for society

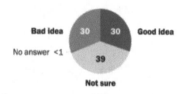

Bad idea 30 | 30 Good idea
No answer <1
39
Not sure

Here's how people were asked to think about gene editing:

"New ways to modify a person's genes are being developed that could make it possible to change the DNA of embryos before a baby is born in order to greatly reduce a baby's risk of developing serious diseases or health conditions over their lifetime."

Note: Figures may not add up to 100% due to rounding.
Source: Survey conducted Nov. 1-7, 2021.
"AI and Human Enhancement: Americans' Openness Is Tempered by a Range of Concerns"

PEW RESEARCH CENTER

There are a large number of potential applications of gene editing techniques for humans.[1] One includes the possibility of using gene editing to prevent, or greatly lower the probability, of developing serious disease. If such applications become widespread it would potentially change the trajectory of human health, greatly reducing the prevalence of serious disease.

The current survey asked respondents to consider one possible use of gene editing techniques: changing the DNA of embryos before a baby is born in order to greatly reduce the baby's risk of developing serious diseases or health conditions over their lifetime.

Among U.S. adults, equal shares (30% each) say the widespread use of gene editing to greatly reduce a baby's risk of developing serious diseases or health conditions over their lifetime would be a good idea or bad idea for society. About four-in-ten (39%) are not sure how they feel about using gene editing for this purpose.

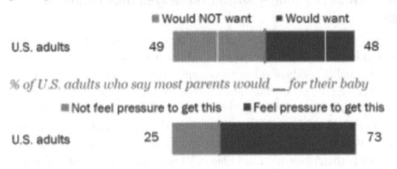

Americans divided over whether they would want gene editing to reduce risk of disease for their own baby

% of U.S. adults who say they definitely or probably ___ gene editing to greatly reduce their baby's risk of developing serious diseases or conditions

■ Would NOT want   ■ Would want

U.S. adults   49   48

% of U.S. adults who say most parents would ___ for their baby

■ Not feel pressure to get this   ■ Feel pressure to get this

U.S. adults   25   73

Note: Respondents who did not give an answer are not shown.
Source: Survey conducted Nov. 1-7, 2021.
"AI and Human Enhancement: Americans' Openness Is Tempered by a Range of Concerns"

PEW RESEARCH CENTER

Americans are about evenly divided over whether they would want to use gene editing in this way for their own baby, if it were available to them. Overall, 48% say they would definitely or probably want this for their own baby; a similar share (49%) say they would not.

Parents of a minor-age child are a bit more hesitant: 42% say they would want this kind of gene editing for their own baby, while 55% say they would not.

Although Americans are closely divided over whether they themselves would want gene editing to reduce the risk of disease for their own baby, a majority thinks that most parents would feel pressure to get this type of gene editing. Nearly three-quarters (73%) think most parents would feel pressure to get gene editing to reduce their baby's risk of developing disease if its use becomes widespread. Far fewer (25%) think most parents would not feel pressure to use gene editing for their baby.

A 2016 Center survey also focused on the idea of using gene editing to enhance health by greatly reducing a baby's chance of developing serious diseases over their lifetime. Americans were also closely divided over whether or not they would want this kind of gene editing for their baby (48% would and 50% would not). However, the current survey and past Center surveys on American's views of gene editing in babies have found large differences in views depending on the intended purpose of the genetic modification.

> **Overall, 53% of U.S. adults say the use of gene editing would be more acceptable to them if it were only used in adults who could consent to the procedure, rather than for babies.**

Concern About Potential Widespread Use of Gene Editing to Reduce a Baby's Health Risk Is Stronger Among Those with High Religious Commitment

One of the largest gaps in views of gene editing to reduce a baby's risk of developing serious disease is between those with higher and lower levels of religious commitment. Those with higher levels of religious commitment (a three-item index based on the importance of religion in their life and their frequency of religious service attendance and prayer) are much more likely to call the widespread use of gene editing in this way a bad idea than a good idea for society (46% to 14%). By contrast, those with low levels of religious commitment are about twice as likely to say it's a good idea for society than to say it's a bad idea (43% to 20%). Between 36% and 41% of those across levels of religious commitment say they aren't sure about their views.

On balance, men are more likely to call the use of gene editing to reduce the risk of disease in babies a good (36%) rather than a bad idea (29%). Women tilt in the other direction, with more saying the widespread use of gene editing for this purpose would be a bad idea than a good one (32% to 24%).

Among those with at least some college experience, views are slightly more positive than negative about the use of this technology. People with a high school diploma or less education are more negative than positive about the implications for society from the widespread use of gene editing in this way.

Highly religious adults more likely to see gene editing to reduce a baby's risk of developing disease as a bad idea than good idea for society

% of U.S. adults who say the widespread use of gene editing to greatly reduce a baby's risk of developing serious diseases or conditions would be a ___ for society

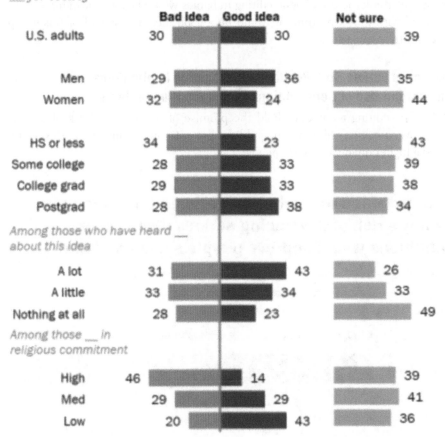

Note: Respondents who did not give an answer are not shown. See Methodology for details on the religious commitment index.
Source: Survey conducted Nov. 1-7, 2021.
"AI and Human Enhancement: Americans' Openness Is Tempered by a Range of Concerns"

PEW RESEARCH CENTER

Just 8% of U.S. adults say they have heard or read a lot about using gene editing to greatly reduce a baby's risk of developing serious diseases or conditions over their lifetime, while 47% say they have heard a little and 44% say they have heard nothing at all about this.

People more familiar with the concept of gene editing for babies to reduce the risk of serious diseases or health conditions during their lifetime are more likely to say it is a good idea than bad idea for society (43% to 31%), while about a quarter say they're not sure (26%). Among those who hadn't heard about this idea prior to the survey, 28% think the use of gene editing in babies would be a bad idea for society, compared with 23% who think it would be a good idea; nearly half of this group (49%) say they aren't sure.

Americans Forsee Both Positive and Negative Implications from Widespread Use of Gene Editing to Enhance Human Health

While gene editing techniques hold the promise of reducing the risk of serious disease over a person's lifetime, the public is not entirely convinced that this would lead to a higher quality of life for people.

Public divided over whether gene editing to reduce a baby's risk of developing serious diseases or conditions would improve people's quality of life

% of U.S. adults who say that if the use of gene editing to greatly reduce a baby's risk of developing serious diseases or conditions becomes widespread, people's quality of life would be ...

■ Better than now   About the same as now   ■ Worse than now

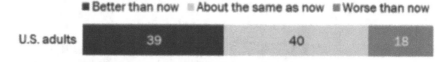

| U.S. adults | 39 | 40 | 18 |

Note: Respondents who did not give an answer are not shown.
Source: Survey conducted Nov. 1-7, 2021.
"AI and Human Enhancement: Americans' Openness Is Tempered by a Range of Concerns"

PEW RESEARCH CENTER

Overall, 39% of U.S. adults think the widespread use of gene editing to greatly reduce a baby's risk of developing serious diseases or health conditions over their lifetime would lead to a better quality of life for people. However, 40% say quality of life for people would be about the same as now if this technology were widely used, and 18% say it would be worse.

Further, about half of Americans (52%) say that "this idea is meddling with nature and crosses a line we should not cross." By contrast, 46% say their views are better described by the statement "as humans, we are always trying to better ourselves and this idea is no different."

Majority of highly religious adults see gene editing to reduce a baby's risk of developing disease as meddling with nature

% of U.S. adults who say that if the use of gene editing to greatly reduce a baby's risk of developing serious diseases or conditions becomes widespread, they would feel that ...

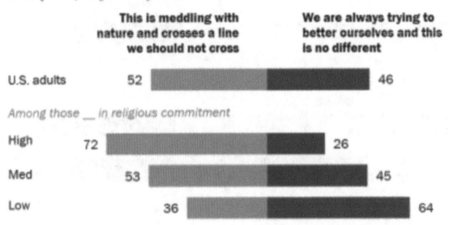

Note: Respondents who did not give an answer are not shown. See Methodology for details on the religious commitment index.
Source: Survey conducted Nov. 1-7, 2021.
"AI and Human Enhancement: Americans' Openness Is Tempered by a Range of Concerns"

PEW RESEARCH CENTER

A majority of those high in religious commitment (72%) consider this use of gene editing to be inappropriate, crossing a line we should not cross. By contrast, a majority of adults with low levels of religious commitment (64%) take the opposing view and describe this use of gene editing as no different than other efforts to better ourselves.

Asked to think about a future where gene editing to reduce the risk of babies developing serious diseases or health conditions is widespread, the public sees both positive and negative impacts as likely to happen. But one negative outcome is seen as particularly likely: the use of these techniques in ways that are morally unacceptable.

Large majority thinks using gene editing to reduce disease risk in babies would lead to some morally unacceptable uses

% of U.S. adults who say that if the use of gene editing to greatly reduce a baby's risk of developing serious diseases or conditions becomes widespread, each of the following definitely or probably ...

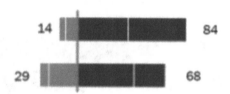

■ Would NOT happen   ■ Would happen

POSSIBLE POSITIVE OUTCOMES

It would lead to new medical advances that benefit society as a whole
31   66

People would live longer and better quality lives
32   65

POSSIBLE NEGATIVE OUTCOMES

Even if used appropriately in some cases, others would use it in morally unacceptable ways
14   84

It would go too far eliminating natural differences between people in society
29   68

Note: Respondents who did not give an answer are not shown.
Source: Survey conducted Nov. 1-7, 2021.
"AI and Human Enhancement: Americans' Openness Is Tempered by a Range of Concerns"

PEW RESEARCH CENTER

Overall, 84% think that even if gene editing is used appropriately in some cases, others would definitely or probably use these techniques in ways that are morally unacceptable.

Nearly seven-in-ten (68%) say these gene editing techniques would definitely or probably go too far eliminating natural differences between people in society.

At the same time, 66% say the development of these techniques would definitely or probably pave the way for new medical advances that benefit society as a whole, and 65% say these techniques would likely help people live longer and better-quality lives.

55% of U.S. adults say widespread use of gene editing to reduce disease risk in babies would lead to more income inequality

% of U.S. adults who say that if use of gene editing to greatly reduce a baby's risk of developing serious diseases or conditions becomes widespread, it would ___ (in) the gap between higher- and lower-income Americans

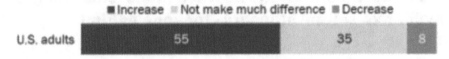

■ Increase ▪ Not make much difference ■ Decrease

| U.S. adults | 55 | 35 | 8 |

Note: Respondents who did not give an answer are not shown.
Source: Survey conducted Nov. 1-7, 2021.
"AI and Human Enhancement: Americans' Openness Is Tempered by a Range of Concerns"

PEW RESEARCH CENTER

In thinking about access to these techniques, 55% of U.S. adults say that if the use of gene editing to reduce a baby's risk of developing a serious disease during their lifetime becomes widespread, the gap between higher- and lower-income Americans would increase. About a third (35%) say the widespread use of this technology would not make much difference in the gap between higher- and lower-income Americans; 8% say it would decrease this gap.

Majority Backs Higher Standards for Testing Use of Gene Editing in Babies, Major Role for Medical Scientists in Setting Standards

Large majority says gene editing techniques should be tested using higher standards than other treatments

% of U.S. adults who say that when it comes to ensuring safety and effectiveness, gene editing to greatly reduce a baby's risk of developing serious diseases or conditions should be tested using ...

■ Existing standards for medical treatments
■ A higher standard than used for medical treatments

| U.S. adults | 17 | 80 |

Note: Respondents who did not give an answer are not shown.
Source: Survey conducted Nov. 1-7, 2021.
"AI and Human Enhancement: Americans' Openness Is Tempered by a Range of Concerns"

PEW RESEARCH CENTER

In line with concerns about the possible misuse of gene editing, a large majority (80%) of Americans say that gene editing techniques to greatly reduce a baby's risk of serious diseases or conditions should be tested using a higher standard than those used for other medical treatments to ensure their safety and effectiveness. Only 17% say gene editing should be tested using existing standards for medical treatments.

Most U.S. adults say medical scientists should play a major role setting standards for gene editing

% of U.S. adults who say each of the following groups should have ___ in setting standards for how gene editing is used

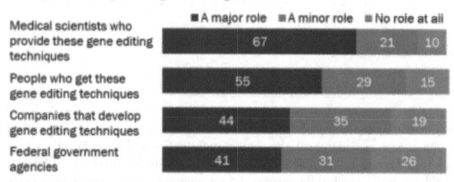

Note: Respondents who did not give an answer are not shown.
Source: Survey conducted Nov. 1-7, 2021.
"AI and Human Enhancement: Americans' Openness Is Tempered by a Range of Concerns"

PEW RESEARCH CENTER

When asked who should be responsible for setting standards regarding the use of gene editing, two-thirds of Americans (67%) believe that medical scientists should play a major role, while another 21% say they should play a minor role.

Over half (55%) say the people who get these gene modifications should play a major role in setting the standards for how they are used; 29% think they should play a minor role.

Fewer than half say the companies that develop gene editing techniques (44%) and federal government agencies (41%) should have major roles in setting standards. However, majorities of Americans say both of these groups should play at least a minor role in setting standards for how gene editing techniques are used.

Public roughly divided over bigger concern regarding government regulation of gene editing techniques

% of U.S. adults who say that if the use of gene editing to greatly reduce a baby's risk of developing serious diseases or conditions becomes widespread, their greater concern is government will ___ regulating its use

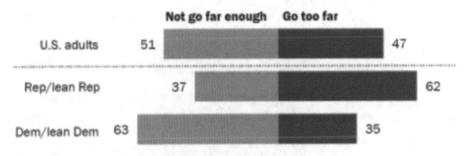

Note: Respondents who did not give an answer are not shown.
Source: Survey conducted Nov. 1-7, 2021.
"AI and Human Enhancement: Americans' Openness Is Tempered by a Range of Concerns"

PEW RESEARCH CENTER

Americans are closely divided when it comes to their greater concern about government regulation in this area. In all, 51% say that if this use of gene editing becomes widespread, their greater concern is government will not go far enough regulating its use. Nearly as many (47%) take the opposite view and say their greater concern is that government regulation will go too far.

Democrats and Republicans differ widely on this question, consistent with their broader views on government regulation. A majority of Republicans and Republican-leaning independents (62%) say that if gene editing to reduce a baby's risk of developing a serious disease or condition becomes widespread, their greater concern is that government will go too far in regulating the use of this technology. Among Democrats and Democratic leaners, 63% say their greater concern is that government regulation will not go far enough.

Consent, Control, and Heritability Matter in Thinking About the Potential Use of Gene Editing to Reduce the Risk of Developing Disease

Overall, 53% of U.S. adults say the use of gene editing would be more acceptable to them if it were only used in adults who could consent to the procedure, rather than for babies; 34% say this wouldn't make a difference in their view and 11% say it would make the use of gene editing less acceptable to them.

About half of U.S. adults say gene editing would be more acceptable under certain scenarios

% of U.S. adults who say each of the following would make the use of gene editing to greatly reduce a baby's risk of developing serious diseases or conditions ...

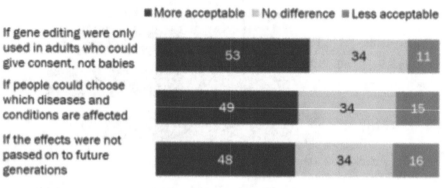

■ More acceptable ■ No difference ■ Less acceptable

If gene editing were only used in adults who could give consent, not babies	53	34	11
If people could choose which diseases and conditions are affected	49	34	15
If the effects were not passed on to future generations	48	34	16

Note: Respondents who did not give an answer are not shown.
Source: Survey conducted Nov. 1-7, 2021.
"AI and Human Enhancement: Americans' Openness Is Tempered by a Range of Concerns"

PEW RESEARCH CENTER

About half or more of both those who say they would and would not want to use gene editing for their own baby say the idea of adult consent for this use of gene editing would make it more acceptable to them.

Roughly half (49%) of Americans also say the use of gene editing to reduce disease risk would be more acceptable to them if people could choose which diseases and conditions are affected. A similar share, 48%, say it would be more acceptable to them if the effects were limited to the person receiving the treatment and not passed on to future generations, a key concern among genetic experts when it comes to the societal and ethical implications of gene editing in babies.

About Seven-in-Ten Americans Favor the Use of Gene Editing to Treat Serious Diseases

When asked to consider some other possible uses for gene editing (beyond use on babies to prevent the future risk of disease) the public is broadly supportive of using gene editing to treat conditions a person already has, but they are broadly opposed to using it to make a baby more attractive.

Majorities favor use of gene editing to treat diseases, oppose its use to enhance a baby's attractiveness

% of U.S. adults who say they would ___ the use of gene editing for the following purposes

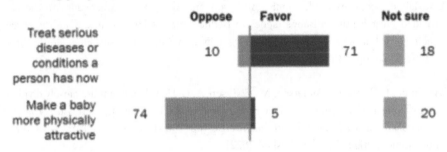

Note. Respondents who did not give an answer are not shown.
Source: Survey conducted Nov. 1-7, 2021.
"AI and Human Enhancement: Americans' Openness Is Tempered by a Range of Concerns"

PEW RESEARCH CENTER

About seven-in-ten Americans (71%) say they would favor the use of gene editing to treat serious diseases or health conditions that a person *currently* has. Just 10% say they would oppose gene therapy, and 18% say they're not sure.

Gene therapy aims to treat disease by correcting an underlying genetic problem. It is in use or in an experimental development phase for a range of diseases including immune deficiencies and cancer.

A 57% majority of those with high levels of religious commitment on a three-item index favor the use of gene editing to treat a current disease or health condition, as do large majorities of those with medium (71%) and low (83%) levels of religious commitment.

By contrast, a large majority (74%) in the U.S. say they would oppose using gene editing to change a baby's physical characteristics to make them more attractive. Only 5% would favor this (20% say they're not sure).

These findings are broadly in line with a 2019-2020 Center survey which found majorities in the U.S. and many other places surveyed thought the use of gene editing in babies to make a baby more intelligent would be taking technology too far but that gene editing for treating a baby's serious disease or health condition would be an appropriate use of the technology.

1. See National Academies of Sciences. 2020. "Heritable Human Genome Editing" for a review of potential applications and cautions about making heritable changes to the human genome.

Print Citations

CMS: Rainie, Lee, Cary Funk, Monica Anderson, and Alec Tyson. "Americans Are Closely Divided Over Editing a Baby's Genes to Reduce Serious Health Risk." In *The Reference Shelf: Gene Editing & Genetic Engineering,* edited by Micah L. Issitt, 136–148. Amenia, NY: Grey House Publishing, 2023.

MLA: Rainie, Lee, Cary Funk, Monica Anderson, and Alec Tyson. "Americans Are Closely Divided Over Editing a Baby's Genes to Reduce Serious Health Risk." *The Reference Shelf: Gene Editing & Genetic Engineering,* edited by Micah L. Issitt, Grey House Publishing, 2023, pp. 136–148.

APA: Rainie, L., Funk, C., Anderson, M., & Tyson, A. (2023). Americans are closely divided over editing a baby's genes to reduce serious health risk. In Micah L. Issitt (Ed.), *The reference shelf: Gene editing & genetic engineering* (pp. 136–148). Amenia, NY: Grey House Publishing. (Original work published 2022)

Bibliography

Ashworth, James. "Dodo 'De-extinction' Announcement Causes Conservation Debate." *Natural History Museum.* Feb. 15, 2023. www.nhm.ac.uk/discover/news/2023/february/dodo-de-extinction-announcement-causes-conservation-debate.html. Accessed Aug. 2023.

Ball, Philip. *Unnatural: The Heretical Idea of Making People.* New York: Penguin Random House, 2012.

Bruening, G., and J. M. Lyons. "The Case of the FLAVR SAVR Tomato." *California Agriculture* 54, no. 4, July 1, 2000.

Chou, Ann F., et al. "Barriers and Strategies to Integrate Medical Genetics and Primary Care in Underserved Populations: A Scoping Review." *Journal of Community Genetics* 12, no. 3, July 2021.

Cohen, Jon. "As Creator of 'CRISPR Babies' Nears Release from Prison, Where Does Embryo Editing Stand?" *Science.* Mar. 21, 2022.

Crichton, Michael. *Jurassic Park.* New York: Arrow Books, 2015.

Dusic, E. J., Tesla Theoryn, Catharine Wang, Elizabeth M. Swisher, and Deborah J. Bowen. "Barriers, Interventions, and Recommendations: Improving the Genetic Testing Landscape." *Frontiers of Digital Health.* Nov. 1, 2022. www.ncbi.nlm.nih.gov/pmc/articles/PMC9665160/. Accessed Aug. 2023.

"Engineered Bacterial Offer a Powerful New Way to Combat Climate Change." *SciTechDaily.* May 10, 2023. scitechdaily.com/engineered-bacteria-offer-a-powerful-new-way-to-combat-climate-change/?expand_article=1.

"Environmental Bioengineering." *MIT School of Bioengineering Sciences and Research.* 2022. mitbio.edu.in/specializations-offered/environmental-engineering/. Accessed Aug. 2023.

"Eugenics and Scientific Racism." *National Human Genome Research Institute/NIH.* www.genome.gov/about-genomics/fact-sheets/Eugenics-and-Scientific-Racism. Accessed Aug. 2023.

Fernández, Clara Rodriguez. "Biotechnology Is Changing How We Make Clothes." *Labiotech.* June 24, 2022. www.labiotech.eu/in-depth/biofabrication-fashion-industry/. Accessed Aug. 2023.

Fliesler, Nancy. "After Decades of Evolution, Gene Therapy Arrives." *Boston Children's Hospital.* Dec. 22, 2020. answers.childrenshospital.org/gene-therapy-history/. Accessed Aug. 2023.

"The Future of Gene Editing." *Columbia University.* Jan. 3, 2020. www.cuimc.columbia.edu/news/future-gene-editing. Accessed Aug. 2023.

"Gene Therapy and Genetic Engineering." *MU School of Medicine*. University of Missouri. medicine.missouri.edu/centers-institutes-labs/health-ethics/faq/gene-therapy#:~:text=The%20distinction%20between%20the%20two,organism%20beyond%20what%20is%20normal. Accessed Aug. 2023.

"Genetic Engineering." *National Human Genome Research Institute/ NIH*. Aug. 31, 2023. www.genome.gov/genetics-glossary/Genetic-Engineering#:~:text=Genetic%20engineering%20(also%20called%20genetic,a%20new%20segment%20of%20DNA. Accessed Aug. 2023.

"Genetic Testing." *CDC*. Genomics & Precision Health. www.cdc.gov/genomics/gtesting/genetic_testing.htm. Accessed Aug. 2023.

Gostimskaya, Irina. "CRISPR-Cas9: A History of Its Discovery and Ethical Considerations of Its Use in Genome Editing." *Biochemistry* 87, no. 8, Aug. 15, 2022. www.ncbi.nlm.nih.gov/pmc/articles/PMC9377665/. Accessed Aug. 26, 2023.

Hallmann, Armin, Annette Rappel, and Manfred Sumper. "Gene Replacement by Homologous Recombination in the Multicellular Green Alga *Volvox carteri*." *PNAS* 94, no. 14, July 8, 1997.

Hinrichs, Katrin. "Controversies About Cloning of Domestic Animals." *Merck Manual*. Nov. 2022. www.merckvetmanual.com/management-and-nutrition/cloning-of-domestic-animals/controversies-about-cloning-of-domestic-animals. Accessed Aug. 2023.

Knoppers, Bartha Maria, and Erika Kleiderman. "'CRISPR Babies': What Does This Mean for Science and Canada?" *Canadian Medical Association Journal*. Jan. 28, 2019.

Langlois, Adèle. "The Global Governance of Human Cloning: The Case of UNESCO." *Humanities and Social Sciences Communications*. No. 3, 2017.

"The Life of Dolly." *Roslin Institute*. Center for Regenerative Medicine. dolly.roslin.ed.ac.uk/facts/the-life-of-dolly/index.html. Accessed Aug. 2023.

Luckhurst, Roger. "An Introduction to *The Island of Dr. Moreau*: Science, Sensation, and Degeneration." *British Library*. May 15, 2014. www.bl.uk/romantics-and-victorians/articles/an-introduction-to-the-island-of-doctor-moreau-science-sensation-and-degeneration. Accessed Aug. 2023.

Mengstie, Misganaw Asmamaw. "Viral Vectors for the *in Vivo* Delivery of CRISPR Components: Advances and Challenges." *Frontiers Bioengineering and Biotechnology*. 2022.

Miko, Ilona. "Gregor Mendel and the Principles of Inheritance." *Nature*. 2008. www.nature.com/scitable/topicpage/gregor-mendel-and-the-principles-of-inheritance-593/. Accessed Aug. 25, 2023.

Mitha, Farhan. "The Return of Gene Therapy." *Labiotech*. Nov. 4, 2020. www.labiotech.eu/in-depth/gene-therapy-history/. Accessed Aug. 25, 2023.

Muacevic, Alexander, and John R. Adler. "Intrauterine Fetal Gene Therapy: Is That the Future and Is That Future Now?" *Cureus*. Feb. 23, 2022.

Murayama, Satomi Angelika. "Op-ed: The Dangers of Cloning." *Berkeley*. May 11, 2020. funginstitute.berkeley.edu/news/op-ed-the-dangers-of-cloning/. Accessed Aug. 2023.

National Research Council. *Genetically Engineered Foods: Approaches to Assessing Unintended Health Effects*. Washington, DC: National Academies of Sciences, 2004.

Pray, Leslie A. "Discovery of DNA Structure and Function: Watson and Crick." *Nature*. 2008. www.nature.com/scitable/topicpage/discovery-of-dna-structure-and-function-watson-397/. Accessed Aug. 25, 2023.

"Questions and Answers about CRISPR." *Broad Institute*. www.broadinstitute.org/what-broad/areas-focus/project-spotlight/questions-and-answers-about-crispr. Accessed Aug. 2023.

Raju, Leela. "Is Blindness Genetic? What to Know." *Medical News Today*. Apr. 11, 2023. www.medicalnewstoday.com/articles/is-blindness-genetic. Accessed Aug. 2023.

Rinde, Meir. "The Death of Jesse Gelsinger, 20 Years Later." *Science History Institute*. June 4, 2019. sciencehistory.org/stories/magazine/the-death-of-jesse-gelsinger-20-years-later/. Accessed Aug. 26, 2023.

"What Is a Gene?" *Medline Plus*. National Library of Medicine. medlineplus.gov/genetics/understanding/basics/gene/. Accessed Aug. 2023.

Worthington, Leah. "De-Extinction Could Reverse Species Loss: But Should We Do It?" *California*. Cal Alumni Association. Sept. 17, 2021. alumni.berkeley.edu/california-magazine/fall-2021/deextinction-revive-ancient-endangered-species-woolly-mammoths-passenger-pigeons/. Accessed Aug. 2023.

Websites

American Society of Human Genetics (ASHG)
Ashg.org

Founded in 1948, the American Society of Human Genetics is a professional organization catering to professionals in human genetics-related fields. The organization also has a number of suborganizations, including the National Society of Genetic Counselors and the American College of Medical Genetics. The ASHG also publishes the *American Journal of Human Genetics* and provides researchers and students with a variety of information about human genetic therapy and research through the organization's website.

Center for Genetics and Society (CGS)
Geneticsandsociety.org

The Center for Genetics and Society (CGS) is a nonprofit organization based out of California that advocates for genetic responsibility and provides informational materials to students on genetics. The CGS has become one of the most active opponents of human cloning and human genetic modification. The organization has also been active in the debate over genetic privacy and over the commercialization of genetic material. The organization has also published numerous articles on genetics controversies in journals and magazines and supports research in genetic ethics.

CRISPR Medicine News
Crisprmedicinenews.com

Danish news agency *Crispr Medicine News* provides articles and information from around the world on the use of Clustered Regularly Interspaced Short Palindromic Repeats (CRISPR) and other genetic editing techniques. The website provides reviews and notices about emerging research, covers books and other media dealing with genetic editing and gene therapy, and supports journalism in the field. Many of the website's resources are provided at no cost for students and researchers seeking information about genetic engineering and the treatment of genetic disease.

Genetic Engineering and Society Center (GESC)
Research.ncsu.edu

The Genetic Engineering and Society Center (GESC) from North Carolina State University provides information about the history, political issues, and cultural debate surrounding genetic engineering and modification. GESC provides numerous free articles and news stories about genetics, gene therapy, and gene editing and also supports and produces original research in the field.

Genetics Society of America (GSA)
genetics-gsa.org

The Genetics Society of America (GSA) is a research and educational organization in operation since the 1930s. The GSA has produced a number of educational and introductory documents about various kinds of genetic engineering and modification and their website provides resources for students and researchers seeking to understand genetics and genetic engineering.

National Human Genome Research Institute (NHGRI)
Genome.org

The National Human Genome Research Institute (NHGRI) is a branch of the National Institutes of Health (NIH) focused on human genetics. The institute publishes and supports research on human genomic issues and supports a program looking into the legal and ethical aspects of genomics research. Free articles available on the NHGRI website provide information about this aspect of the genetics debate.

Wired
Wired.com

Wired is a monthly magazine ongoing since 1993 that focuses on culture, technology, economics, and politics. The magazine consistently covers new developments in the field of genetic engineering and offers updates on ongoing gene research programs. Written for the general audience, *Wired* provides general readers with introductions to a variety of genetics topics, though access to all articles requires a yearly subscription.

Index